WISCONSIN'S FOUNDATIONS

*A Review of the State's Geology and Its
Influence on Geography and Human Activity*

Gwen M. Schultz

Associate Professor of Geography
The University of Wisconsin System

With a new foreword by
James M. Robertson

*Cooperative Extension Service
University of Wisconsin*

The University of Wisconsin Press

The University of Wisconsin Press
1930 Monroe Street, 3rd Floor
Madison, Wisconsin 53711-2059
uwpress.wisc.edu

3 Henrietta Street
London WC2E 8LU, England
eurospanbookstore.com

Printed in the United States of America

Library of Congress Cataloging-in-Publication Data
Schultz, Gwen.
Wisconsin's foundations : a review of the state's geology and its
influence on geography and human activity / Gwen M. Schultz ; with a
new forward by James M. Robertson.
    p. cm.
    Originally published: Dubuque, Iowa : Kendall/Hunt Pub. Co.,
© 1986.
    Includes bibliographical references and index.
    ISBN 0-299-19874-X (pbk. : alk. paper)
    1. Geology—Wisconsin. 2. Human geography—Wisconsin.
I. Title
QE179.S38   2004
557.75-dc22   2003070543

ISBN 13   978-0-299-19874-9 (pbk. : alk. paper)

Dedicated to

Professor Lawrence G. (Larry) Monthey

environmentalist and educator

in recognition and remembrance of his valuable contribution
to the people of Wisconsin

(1918–1985)

Contents

Contents

Foreword

Most Wisconsin citizens share a deep appreciation of the shape and texture of their familiar landscapes, the abundance of fresh water, the fertile soils, the northern forests, and the lack of life-threatening natural hazards that characterize our state. All these features are directly related to a special set of geologic processes and materials that collectively define the land on which we all live, work, and play. It is natural to be curious about the history of such a pleasant place—How did it come to be this way? How did it look in the past? What kinds of creatures lived here before us? In Wisconsin's case, the geologic story is long, complex, and incomplete, beginning over three billion years ago and still in progress.

Bits and pieces of this story have been told in the scientific literature for years. But the nonspecialist had no easy way to discover the fascinating stories of Wisconsin's past until Gwen Schultz prepared this book in 1986. Her synthesis of the state's geology, geography, and human development filled a special need: It effectively reached out to a broad audience of interested citizens who simply want to know more about the origins, evolution, and geological underpinnings of the Wisconsin landscape.

At the Wisconsin Geological and Natural History Survey, I see on an almost daily basis the hunger that the public has for the kind of understandable information Gwen has supplied in *Wisconsin's Foundations.* I know of no other book that meets this particular need for our state. I am grateful that the University of Wisconsin Press decided to reprint this long out-of-print book.

James M. Robertson
Director and State Geologist
Wisconsin Geological and
Natural History Survey

Preface

For a long time there has been a need for a well-rounded, comprehensive book on the geology of Wisconsin, one the average person can understand and use. Many years have passed since the State last commissioned a book to be written on this subject, and meanwhile the lack has been seriously felt by educators, workers in many other fields, and the general public. Therefore, the University of Wisconsin–Extension, which is tuned to the people's needs, responded and initiated the writing of this book.

In the past, two previous State-sponsored publications have been heavily relied upon for geologic information about Wisconsin. The first was T. C. Chamberlin's four-volume *Geology of Wisconsin* published over a century ago. That detailed, monumental report, by Chamberlin and a team of other men, was based on the 1873–1879 survey of Wisconsin. Lawrence Martin's more concise and less technical *Physical Geography of Wisconsin* was published in 1916 and slightly revised in 1932.

Much of the material published about Wisconsin geology over the years has not been known or readily accessible to non-professional researchers and readers—that in scholarly journals, monographs, field-trip guides, reports of professional societies, topical reference maps, and so on. Many of those publications use terminology and means of presentation that are foreign to people untrained in geology and related fields. Also, most of them deal with just a certain locality, a limited subject, or research problems.

What was wanted here was a screening and synthesis of significant, generally accepted facts and concepts; a pulling together of information from many sources; and a condensed telling of the composite geologic story of the whole state in language understandable to a non-specialist. For having been offered the opportunity to write this current book about her home state, the author is indeed grateful.

It is written for average Wisconsin residents, students, environmentalists, public workers, visitors—for all interested people, within and outside Wisconsin. Its purpose is to describe and explain Wisconsin's geology, and secondarily to show, within book-length limitations, how geologic conditions have affected people's lives and activities. Thus the word "foundations" in the title has two meanings. One is the physical base of rocks and earth materials. The other is the basis (in part) of the area's historical and economic development, and of its cultural aspect and land use.

Geologic terms that were considered overly cumbersome for the lay reader, or whose definition or usage is not universally agreed upon, have been avoided where possible. Preference has been shown for terms that are standard in earth-science teaching and literature, that are in common use in our area, and that are not subject to continual redefinition. Special attention has been given to explaining older terms that were formerly popular in historic geologic writings and maps in order to aid in the understanding and "translating" of those still-valuable sources of information. Allusions are made to older literature now and then, partly as a reminder that some of the best recording and description is found in the works of scholars of past generations. Old photographs are included, too, to give a sense of the past, and because the intention is not to depict just a moment of present time, but to show also how features appeared in their natural state, how they came to be what they are, and how landscapes looked before they were altered. As for recent photos, an effort was made to obtain them not only from professional photographic sources, but from local field workers and other people throughout the state as well, so as to have wide representation from many contributors.

Certain parts of the text may seem elementary to an advanced geology scholar, but are needed to provide background for the less-informed reader. On the other hand, some rather technical material had to be included. But controversial and speculative topics are not discussed in depth, though many are mentioned, as this is not the place to evaluate currently debated theories and opinions. The future will say who is right. Many concepts now held in geology, as in other sciences, will eventually change. New instruments and techniques bring new discoveries, and gradually fresh perceptions are gained of how, and when, things happened in the past. For now, this is a compressed summary of highlights of Wisconsin's geologic makeup and the chronology of events that made it so, as perceived and compiled by one writer at this time.

Gwen M. Schultz
Madison, Wisconsin

Acknowledgments

The author expresses her sincere appreciation to the administrators of the University of Wisconsin–Extension for their support of this book project—their financial backing, their patience, and their help in many ways. She especially wishes to thank Dr. Gale L. VandeBerg, retired Dean/Director of Cooperative Extension, for asking her to write the book. Many others in the University community are thanked too for their encouragement and assistance.

The following should be specifically acknowledged for their help.

The Wisconsin Geological and Natural History Survey for its technical support, and its staff members who critically commented on parts or all of the manuscript and supplied data.

Other manuscript reviewers or providers of information, including professors J. Campbell Craddock, Robert H. Dott, Jr., and David Mickelson of the University of Wisconsin–Madison Geology Department; and geology professors Gene LaBerge of UW–Oshkosh, Paul Myers of UW–Eau Claire, and Paul Tychsen of UW–Superior. Also the following geography professors: Adam Cahow of UW–Eau Claire, Robert Finley of UW–Extension, James Knox of UW–Madison, Gene Musolf of UW–Marathon County, and Ray Pfleger and Dan Zielinski of UW–Waukesha.

Carl Dutton of the U.S. Geological Survey for generously giving his counsel and knowledgeable insights into Precambrian geology.

Klaus W. Westphal, Curator of the Geology Museum of UW–Madison for selecting and identifying fossil and rock specimens, and supervising the photography and plate preparation for illustrations.

John Dallman, Curator of Paleontology, Department of Zoology, UW–Madison, for his contributions in the realm of Pleistocene and other fauna.

Staff members of the State Historical Society of Wisconsin—Joan Freeman, John Penman, Howard Kanetzke, and Doris Platt.

Ken Dowling and David Engleson of the Wisconsin Department of Public Instruction; and Lola Pierstorff, former Director of Instructional Materials, Department of Education, UW–Madison.

Ruth Hine and George Knudsen of the Wisconsin Department of Natural Resources.

Many librarians, especially Mary Galneder of the Arthur H. Robinson Map and Air Photo Library; Miriam Kerndt of the Geography Library in Science Hall; and the staff of the Geology Library, Weeks Hall.

Professor Lawrence (Larry) G. Monthey of the UW Environmental Resources Center for generously sharing his familiarity with the natural scene, and his great fund of knowledge about the state, its significant sites and points of interest.

And others who in special ways contributed to the book's content and completion—Professor Jerry Culver, Department of Geography, UW–La Crosse; Emeritus Professor Cotton Mather, University of Minnesota, Minneapolis; Emeritus Professor Randall Sale, cartographer, UW–Madison Geography Department; and Captain Charles A. Widmann, United States Navy Retired.

The staff of the university's Photographic Media Center for their care and skill in preparing photographs.

Mapping Specialists, Ltd., of Madison, were the book's cartographers.

A Synopsis of Wisconsin's Geologic History

Wisconsin's geologic landscape, like a landscape anywhere, invites the perceptive observer to study it and learn how it formed and how it changed through time's passing eras.

One who merely scans Wisconsin's gently contoured, stable terrain without studying it can wrongly surmise that its geologic past was relatively simple and uneventful. The landforms of modest proportions; the lack of mountains and other grandiose rock structures; the surficial smoothing and camouflaging done by glaciation: this visual plainness is misleading. Wisconsin's geologic story has been as complex and spectacular as that of any place on Earth. Here in Wisconsin are impressive records of rock rending, volcanic outpourings, flooding seas, and grinding ice sheets; and here are found some of the oldest rocks, and some of the most diverse associations of rocks, on the planet.

While the landscape's origin and composition alone make an interesting study, the land's greater significance is the role it plays in people's lives. Rocks, waters and soils model the physical setting and are the platform upon which the human drama of prehistory and history has been enacted. They have a strong influence on how people live and what they can do. Conversely, people have a small but ever-increasing impact upon the geologic landscape in the way they utilize and remodel it. Their effect may be slight, superficial and transitory in the long, cosmic view of things, but it is all-important to the resource base and the land's esthetic quality during this period of human occupation, however long that will be. It is well to keep in mind how these two factors—the region's geologic foundations; and the people who reside, work and travel upon them—interact with each other.

(The reader will be helped by referring to the plates at the end of the book and comparing them with each other as appropriate.)

Wisconsin's Setting

Wisconsin resembles a left hand set down near the middle of North America, a little east of center, approximately between latitudes 42½° and 47°N (plate 6). Its thumb, Door Peninsula, claims Lake Michigan. Its north-reaching fingers join the continent's ancient core. Its palm area consists of a layered assortment of middle-aged rocks overlapping those older rocks. Its wrist merges into the broad interior lowland of the continent. Covering most of the back of that hand is a glove of glacial deposits.

Wisconsin's irregular boundaries are determined in large part by noteworthy features. If you enter Wisconsin from the west you cross the Mississippi River, the continent's greatest river. It and a tributary, the St. Croix, form most of the state's western boundary. Along that boundary one can image the profile of an Indian's face, looking west. Its outline helps in designating locations, and it gives the name "Indian Head Country" to the northwestern part of the state. If you enter Wisconsin from the north you cross Lake Superior, largest of the Great Lakes and, in area, the world's largest fresh-water lake. Enter from the east and you cross Lake Michigan, somewhat smaller but still one of the world's largest lakes. In the northeast Wisconsin is separated from Michigan partly by an artificial boundary and partly by the Montreal River and the Menominee River, with its tributary, the Brule. (This river should not be confused with the Bois Brule River in eastern Douglas County, which is commonly called just "the Brule.") In the south Wisconsin's fertile farmland continues without interruption into Illinois.

The area of Wisconsin is 56,154 square miles (146,000 sq km). This area figure, the one generally used, includes inland waters but does not include the parts of lakes Michigan and Superior that belong to Wisconsin. Official state boundary lines run not along the shores of the Great Lakes, but rather through the lakes, dividing them among the bordering states and Canada. The portions of lakes Michigan and Superior lying within Wisconsin's boundaries total 6,439,700 acres, more than 10,000 square miles.

Geologic events that built and shaped Wisconsin took place on a broad stage that reached far beyond this state's boundaries, of course. In this synopsis our view will expand to continental and global scenes when it is necessary to give perspective to what was happening in Wisconsin.

The Beginnings

Wisconsin's geologic beginnings are obscured in the blurry genesis of Earth itself. It is believed that when the young planet was in its formative stages it was somewhat molten, and that as its outer crust was developing relatively light molten material worked its way toward the surface in places and hardened, while heavier material gravitated toward the planet's center. Masses of less dense rock that aggregated at the surface presumably stood higher than the rest of the crust and may have constituted areas where continents started to take form. In low areas water collected as oceans of unknown size and depth.

The raw, young landmasses rising above the oceans were still without vegetation. They must have been subjected to rapid weathering (decay and disintegration) and to vigorous erosion, or wearing away, as rains fell upon bare rock; but apparently they continued to widen as additional molten material welled up from below and built outward into the oceans, and as rock debris washed down from their own surfaces.

While the landmasses were forming, some of them—and maybe all—were not in the locations where they are now. Supercontinents formed, then split into smaller continents which slowly moved away from each other over millions of years as ocean floors widened between them. Before they split apart, North America, Greenland and Eurasia were united, as were the southern-hemisphere continents (fig. 1.1). This moving of continents is often called simply "continental drift," but it involves complicated processes of sea-floor spreading and plate tectonics, whereby plates, or sections of the earth's crust, slowly move. Drifting landmasses not only divided, but sometimes collided slowly with others and joined, causing deformation of affected rock, especially along the margin where contact occurred. Through these processes of splitting and recombining, and volcanic upwelling of new material from within the earth, the present continents took shape. These processes still go on, and the shape of continents gradually continues to change.

During early geologic time the incipient continents were reworked by volcanism, by bending and breaking

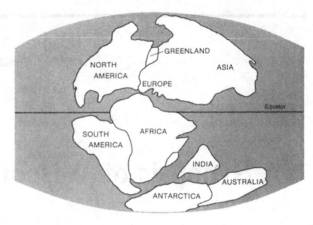

Figure 1.1 A generalized concept of how main landmasses may have been positioned about 180 million years ago as they were dividing and moving. (After R. S. Dietz and J. C. Holden, *Journal of Geophysical Research*)

of rock, by erosion and redeposition of materials, by dipping into the sea and rising, and by splitting apart and rejoining in different ways until they were forged and welded into exceptionally resistant, stable landmasses. Their mountains and other highlands were worn down until most of their surfaces were smoothed to plains and rounded hills, with higher prominences left only here and there. These old, contorted, reshaped, long-eroded landmasses were the cores and bases around and upon which new rocks would form, filling out the shapes of future continents. Wisconsin's oldest foundation rock is part of the southern slope of one of these ancient stable landmasses, which was a nucleus of what ultimately would become North America.

Each present continent contains at least one such ancient core. These cores are termed *shields* where they appear at the surface over wide areas. They bulge up toward the center like medieval battle shields, and the term connotes hardness and strength. North America's shield is called the *Canadian Shield,* since it comprises the eastern two-thirds of Canada. It is sometimes called the Laurentian Shield or the Laurentian Upland because its rocks were first studied in the Laurentian Hills along the St. Lawrence River.

The Canadian Shield has a conspicuous "dent" in it, the depressed Hudson Bay area. On maps the shield is seen encircling the bay, including most of central and eastern Canada with parts of the arctic islands, and most of Greenland, and extending south to take in a section of the United States—the northern parts of Minnesota, Wisconsin, Michigan, and New York (fig. 1.2). Peripheral parts of the shield are buried by younger rocks and extend out under them as "basement rock" for unknown distances—hundreds of miles in some directions—and at varying depths. In places the surface of the buried shield is known to be thousands of feet underground.

Figure 1.2 The Canadian Shield of North America.

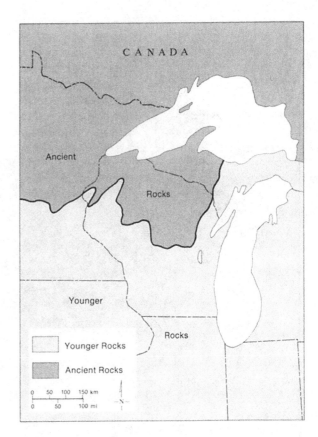

Figure 1.3 Generalized diagram showing Wisconsin as a transitional zone between ancient and younger rocks.

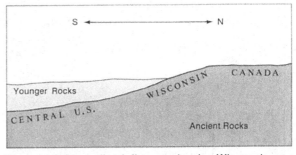

The shield surfaces above younger rocks only in the northern part of Wisconsin, but its buried extension underlies all the rest of the state, being the base upon which the younger rock formations later accumulated (fig. 1.3). The depth of the shield's uneven buried surface is known because some river valleys have been cut deep enough to expose it, well borings have encountered it, subsurface soundings with sensitive instruments have located it, and here and there hills which were left on its otherwise leveled surface show their summits through the covering of younger rock and glacial deposits. The best known of these protruding hill features within the state is the Baraboo Hills, or Baraboo Range, in southcentral Wisconsin in Sauk and Columbia counties (fig. 1.4).

The highest part of the shield in the state, and the highest part of the state, is northcentral Wisconsin. From there the shield's buried surface slopes downward in all directions—steeply toward the east and southeast, and more gently to the west. In the southeastern corner of the state it pitches down to a depth of about 2,000 feet (600 m) below sea level. (The surface level of Lake Michigan now is about 580 feet [177 m] above sea level.) Along the middle section of Wisconsin's Lake Michigan shore, in most of Door Peninsula, and also in the extreme southwestern corner of the state the shield is at a shallower depth, but still more than a thousand feet below sea level. Along the middle section of the Mississippi River border its depth is a little above sea level.

Wisconsin's geological portrait is enhanced by containing part of the Canadian Shield in its composition. Rocks of the shield in northern Wisconsin are generally concealed by glacial deposits, but locally they are exposed—that is, not covered. Their composition and structure, and their relationship to shield rocks in neighboring areas, reveal or hint at exciting episodes in the development of the Wisconsin area and of this planet as far back as three billion and more years ago.

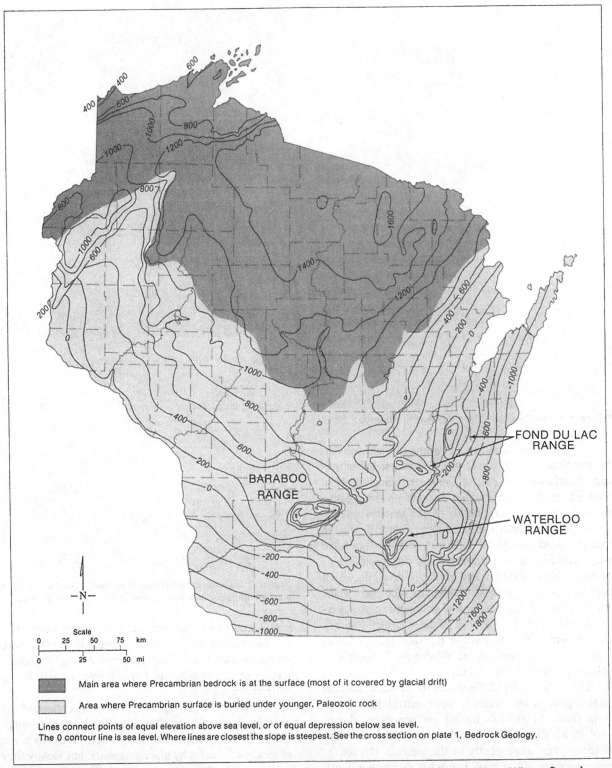

Figure 1.4 The Precambrian Surface. (After F. T. Thwaites and others. Wisconsin Geological and Natural History Survey)

Main area where Precambrian bedrock is at the surface (most of it covered by glacial drift)

Area where Precambrian surface is buried under younger, Paleozoic rock

Lines connect points of equal elevation above sea level, or of equal depression below sea level.
The 0 contour line is sea level. Where lines are closest the slope is steepest. See the cross section on plate 1, Bedrock Geology.

FOND DU LAC RANGE

WATERLOO RANGE

BARABOO RANGE

Scale
0   25   50   75   km
0   25   50   mi

Figure 1.5 Examples of igneous rocks: **(a)** A fine-grained granite. This specimen is about 4 inches, or 10 centimeters, wide. (Ralph V. Boyer) **(b)** A coarser-grained granite. This specimen is 5 inches wide. (Gwen Schultz) **(c)** Gabbro, a coarse-grained igneous rock, from northeast of Mellen, Ashland County. (Ralph V. Boyer) **(d)** An old chopping tool fashioned of basalt, a fine-grained, hard igneous rock. (Ralph V. Boyer) **(e)** Basalt (sometimes called trap rock) which formed from lava, as exposed in a natural setting in western Polk County. (George Knudsen, Wisconsin Department of Natural Resources)

How the Rocks Formed

The three general classes of rock—igneous, sedimentary and metamorphic—are all well represented in Wisconsin, in abundance and variety. The ancient shield is mainly a complex of igneous and metamorphic rock, with some areas of sedimentary rock. The younger rocks of the state are predominantly sedimentary.

IGNEOUS rocks are those that formed from molten material which solidified when it cooled. It may have been lava or other volcanic material that poured or erupted onto the earth's surface, forming rocks such as basalt or rhyolite; or it may have been magma that cooled more slowly below the surface, forming rocks like granite or gabbro (fig. 1.5). The first rocks to form on Earth must have been igneous, but once weathering attacked those that were exposed, causing them to break apart and decompose, the materials necessary for the making of sedimentary rock were available.

SEDIMENTARY rocks are those made of materials eroded or dissolved from pre-existing rocks—materials carried mainly by streams,[1] and deposited or precipitated in low places where they could accumulate and remain undisturbed. The largest areas of accumulation have been, and still are, the gently sloping, submerged, nearshore margins of the continents where streams spill their loads. There along shorelines in shallow seas rock particles are sorted by waves and currents, and those of like weight drop out together in different zones. The coarsest, heaviest particles come to rest first, nearest the shore. The finest, lightest particles, which stay suspended longest, can be carried farthest to settle out last in quiet, usually deeper water. Dissolved material is chemically precipitated (as lime is deposited in water containers). After long compres-

1. A stream, in geologic language, is any body of flowing water, large or small, from a mighty river to a trickling rill.

Figure 1.6 Examples of sedimentary rocks: **(a)** Dolomite, a resistant type of limestone, in Peninsula State Park, Door Peninsula. Its horizontal bedding is apparent, as are numerous joints, or cracks, which run vertically through the rock in this eroding outcrop. (George Knudsen, Wisconsin Department of Natural Resources). **(b)** Conglomerate rock composed mainly of fragments of quartzite, which were deposited at Weidman Falls at the southwest end of the Baraboo Hills during Late Cambrian time. Scale line = 2 cm (0.8 in.). (Geology Museum, University of Wisconsin-Madison. Photographer, Lawrence D. Lynch) **(c)** A conglomerate of larger components. The largest boulders here are about a foot across. The location is Parfrey's Glen, also in the Baraboo Hills. (Ralph V. Boyer) See sedimentary rocks also in Fig. 1.15 and chapters 4 and 5.

sion under the weight of heavy sediments later laid down above, and after cementation by substances carried in percolating groundwater, the beds of deposited materials harden into layers of sedimentary rock. If the rocks are uplifted or the sea withdraws, the rock emerges as land. Many sedimentary rocks contain fossilized remains of plant and animal life.

Different kinds of deposited materials formed (and are still forming) different kinds of sedimentary rocks (figs. 1.6, 1.14 and 1.15):

Shale, which resembles (and originally was) compact, dried mud, formed from fine silts and clay particles that settled out in quiet water.

Sandstone formed from sand grains which were battered, ground, and sorted by water, and sometimes by wind.

Conglomerate formed from a mixture of unsorted fine and coarse materials of the kind found along shores. It is a conglomeration of materials of various sizes, which may include pebbles, cobbles or boulders held in a cementlike matrix of finer rock particles. Conglomerate often resembles concrete.

Limestone formed from lime precipitates released from the water, and also from shells, corals, and other calcareous (composed of calcium carbonate) remains of sea life. These materials accumulated in somewhat deeper water, or in other water where there was not much sand and silt.

In the Wisconsin area most rocks deposited originally as limestone were changed to *dolomite* by the chemical replacement of calcium by water-borne magnesium. Dolomite has also been called "magnesian limestone," and some geologists prefer to call it "dolostone" because, strictly speaking, "dolomite" is a mineral which the rock called dolomite approximates in composition.

Fragments of sedimentary earth material range in size, from smallest to largest, from the finest particles of clay, to silts, sands, pebbles, cobbles, and the largest—boulders. One commonly used scale classifies pieces of rock sediment by size:

Clay—from the smallest possible size up to silt size

Silt—1/256 mm to 1/16 mm diameter (fine powder)

Sand—1/16 to 2 mm diameter

Pebble—2 mm to 64 mm diameter

Cobble—64 mm to 256 mm diameter

Boulder—larger than cobble

C. K. Wentworth, After J. A. Udden

Gravel, in geologic terminology, is an accumulation of smoothed, rounded rock pieces of pebble size and larger. The smoothing was done mainly by water as it agitated the rock pieces, and in some cases by wind or glacial ice.

METAMORPHIC rocks result when igneous or sedimentary rocks become altered, or metamorphosed,

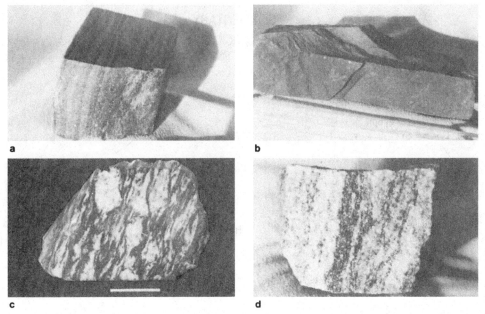

a   b

c   d

Figure 1.7 Examples of metamorphic rocks: **(a)** Baraboo Quartzite from the Baraboo Hills at Devil's Lake. The upper side of the 4-inch specimen has been polished. It shows the layering of the original sandstone which was strongly cemented and metamorphosed. (Ralph V. Boyer) **(b)** Barron Quartzite from western Sawyer County. (Ralph V. Boyer) **(c)** Precambrian gneiss from near Neillsville. It illustrates how, through pressure and strains of metamorphism, unlike minerals of a preexisting rock were stretched out, forming parallel bands of lighter and darker color which typify this rock. (More specifically, this specimen is called "augen gneiss" because some mineral fragments remain unflattened or only partly flattened, resembling eyes. "Augen" means "eyes" in German.) Scale line = 3 cm (1.2 in.). (Geology Museum, University of Wisconsin-Madison. Photographer, Lawrence D. Lynch) **(d)** This piece of gneiss, which is about 5 inches across, is from Mill Creek southwest of Stevens Point and is said to be about 1,940,000,000 years old. (Ralph V. Boyer)

by intense heat, pressure or chemical action (fig. 1.7). So sedimentary sandstone can turn into metamorphic quartzite. Shale can become slate. Limestone or dolomite metamorphosed can become marble. Igneous granite can become banded gneiss. A metamorphic rock itself can be altered again and so become a different kind of metamorphic rock. Any kind of rock can be "recycled" over a period of time; through fragmentation it can become material for new sedimentary rock, or through reheating it can become new igneous rock (fig. 1.8).

Igneous and metamorphic rocks together are sometimes referred to as crystalline rocks, since most consist mainly of interlocking crystals.

The Progression of Time and the Age of Rocks

The Canadian Shield, like the world's other shields, formed during that long, distant era geologists call Precambrian (fig. 1.9). As its name signifies, the Precambrian Era spanned all of geologic time before the Cambrian Period—nearly 90 percent of Earth time. The Precambrian started at Earth's beginning and ended at a hazy time boundary estimated to have been around 600 million years ago. Relatively little is known about what happened in the Precambrian Era, especially during its earliest phase. The farther back in time one looks, the more the rock evidence has been eroded

away, the more the remaining rocks—all we have to go by—have been metamorphosed, and the more obscure the clues to their sources and development have become.

Although the boundary between the Precambrian and the Cambrian is unclear and transitional, it has a special significance. When a geologic time chart was first constructed in the nineteenth century the belief was that no life existed on Earth before the Cambrian, that not until the Cambrian did the first simple life forms appear in the oceans. Fossils (naturally preserved remains of once-living things) had not yet been found in any Precambrian rocks, but they were abundant in those of the Cambrian Period. Later, geologists learned that the Precambrian had indeed contained life, marine creatures; but they were either so small (the earliest were one-celled organisms) or their remains were so soft (they had no skeletons, shells, or other hard parts) that few fossils are preserved.

In Cambrian time and thereafter, life forms were numerous enough, and evidences of their existence enduring enough, to provide a nearly continuous chronological record. On ocean floors around the world fossilized remains and imprints were left in soft, undisturbed sediments. Those that were covered, mineralized, and protected from destruction were able to endure through many millions of years as the sediments hardened into rock. In sedimentary rocks formed

7

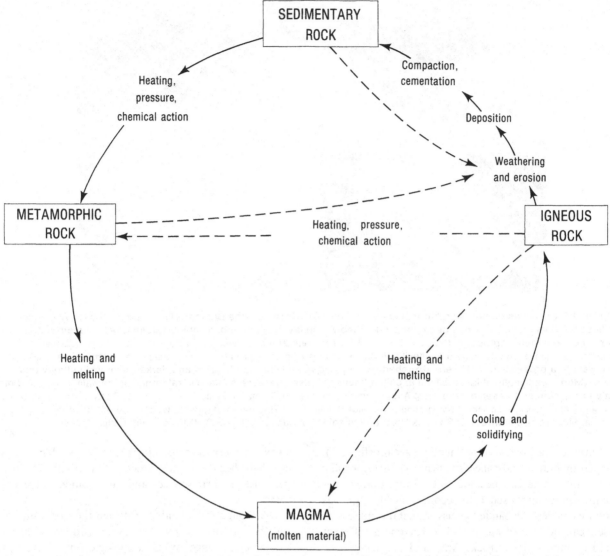

Figure 1.8 The rock cycle. Rocks slowly change from one form to another. Rock material can follow any of the arrows from one phase to another.

during and after Cambrian time, an orderly sequence of fossils of animal and plant life has proved to be an invaluable key to time intervals and to the ages of rocks. Sediments were deposited in layers, or strata, one atop the other; the oldest layer of a sequence is, naturally, at the bottom, and consecutively younger layers lie above it. The community of fossils in a given layer tells where it fits in the time scale, since the escalator of evolution, with its ever-changing assemblage of passengers, travels in only one direction.

Although fossils are valuable indicators of the age of sedimentary rocks, there is a limit to their usefulness in dating rocks; and there are perplexing problems of matching and correlating strata in separated places, and of bridging gaps in the sequences of strata. Fossils tell only the comparative or relative ages, or order of deposition, of given rock strata. They cannot tell, just by their presence, how long ago it was that they died and

the sediments were deposited around them. That is, fossils in rock do not give absolute numerical dates. Besides, sedimentary rocks, which contain the preponderance of fossils, make up only part of the earth's surface. Most of the other rocks, the igneous and the metamorphic, contain no fossils. Intense heat associated with the forming of igneous rocks precluded life; and great pressure, heat, and distortion during the metamorphism of sedimentary rocks mutilated most fossils, if, indeed, fossils had existed at all. The very ancient sedimentary rocks, those that formed before life appeared, have no fossils, of course; and even many younger rocks lack them.

Fortunately other methods of determining the age of rock have been developed and are helpful in the dating of ancient and complex rocks. Radiometric techniques can be used to measure the degree of disintegration of certain unstable radioactive elements in

8

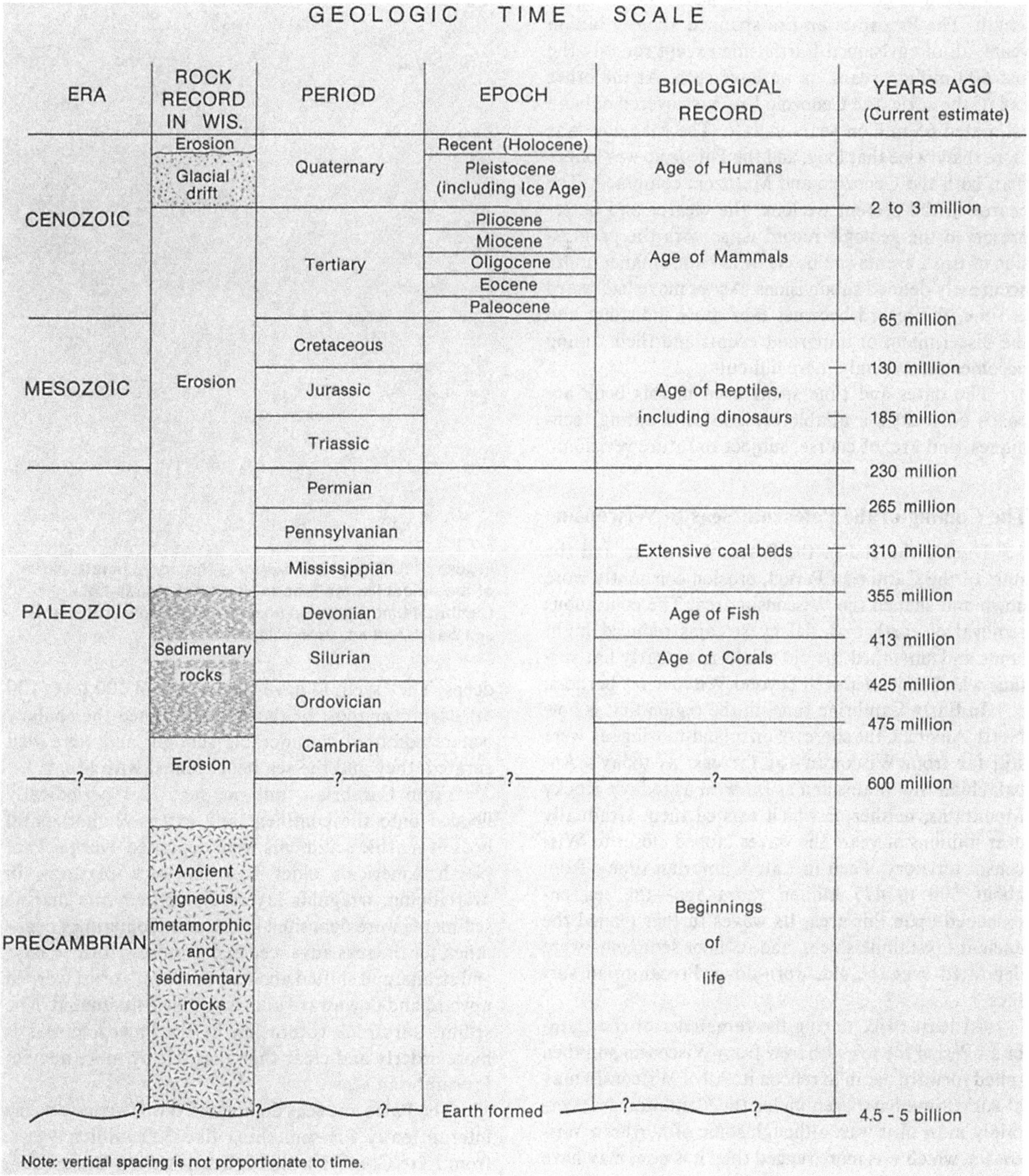

GEOLOGIC TIME SCALE

ERA	ROCK RECORD IN WIS.	PERIOD	EPOCH	BIOLOGICAL RECORD	YEARS AGO (Current estimate)
CENOZOIC	Erosion	Quaternary	Recent (Holocene)	Age of Humans	
	Glacial drift		Pleistocene (including Ice Age)		
		Tertiary	Pliocene	Age of Mammals	— 2 to 3 million
			Miocene		
			Oligocene		
			Eocene		
			Paleocene		
MESOZOIC	Erosion	Cretaceous		Age of Reptiles, including dinosaurs	— 65 million
		Jurassic			— 130 million
		Triassic			— 185 million
PALEOZOIC		Permian			— 230 million
		Pennsylvanian		Extensive coal beds	— 265 million
		Mississippian			— 310 million
	Sedimentary rocks	Devonian		Age of Fish	— 355 million
		Silurian		Age of Corals	— 413 million
		Ordovician			— 425 million
		Cambrian			— 475 million
——?——	Erosion	——?——	——?——	——?——	— 600 million
PRECAMBRIAN	Ancient igneous, metamorphic and sedimentary rocks			Beginnings of life	
——?——	——?——	——?——	Earth formed ——?——	——?——	— 4.5 - 5 billion

Note: vertical spacing is not proportionate to time.

Figure 1.9 Geologic time scale.

a rock, elements that decay at a known slow, constant rate, and thus give an estimate not only of the relative age of the rock in which they occur (as fossils do), but also of the absolute numerical age in years. As the techniques are perfected and more widely applied, knowledge of the ages of rocks and of the timing of geologic events should increase greatly.

Between the Precambrian-Cambrian boundary and the Present three main time zones are defined, based on the fossil record. From oldest to youngest they are:

the Paleozoic (*paleo* = old; *zo* = life), the era of ancient life;

the Mesozoic (*meso* = middle), the era of middle life:

the Cenozoic (*ceno* = recent), the era of recent life, which includes the time in which we live.

The major geologic time divisions are not of equal length. The Precambrian Era spanned about 4 billion years, all of envisioned Earth time except roughly the last 600 million years, as we have seen. At the other end of the scale, the Cenozoic Era has covered only an estimated 65 million years to date. The Mesozoic was more than twice that long, and the Paleozoic was longer than both the Cenozoic and Mesozoic combined. The nearer to the present we look, the clearer and better preserved the geologic record is; so with the progression of time, events can be classified into smaller, more accurately defined subdivisions. As we move backward in time, the record becomes ever more indistinct and the discernment of important events and their timing becomes increasingly more difficult.

The dates and time spans used in this book are based on today's available evidence and dating techniques, and are, of course, subject to future revision.

The Coming of the Paleozoic Seas to Wisconsin

During the end of the Precambrian Era and the start of the Cambrian Period, erosion constantly wore down and shaped the Wisconsin area. The continuous removal of earth material by streams reduced highlands and smoothed the old shield to a nearly flat surface which extended well beyond Wisconsin's borders.

In Early Cambrian time, in the region that is now North America, the shores of an inland-moving sea were still far from Wisconsin—as far east as today's Appalachian Mountains and as far west as today's Rocky Mountains, neither of which existed then. Gradually over millions of years the waves lapped closer to Wisconsin territory. Then in Late Cambrian time—from about 500 to 475 million years ago—the sea encroached upon this area. Its waves further planed the ancient crystalline shield, and its floor sediments were deposited over the old, worn-down Precambrian surface.

At least twice during the remainder of the Cambrian Period the sea withdrew from Wisconsin and then rolled forward again to reflood it. All of Wisconsin may at some time have been under the Cambrian sea; certainly most of it was, although some of northern Wisconsin, which was more rugged than it is now, may have remained above water.

The Cambrian sea encircled the whole Canadian Shield, and during its maximum encroachment probably left little of the shield core above water (fig. 1.10). It, like all of Wisconsin's seas during the Paleozoic, is descriptively termed an "epicontinental sea" (epi = upon); that is, a shallow body of water lying upon the continent as distinguished from the much deeper waters in ocean basins. Such seas resembled the waters over continental shelves, like those off southeastern United States, rather than those of the ocean

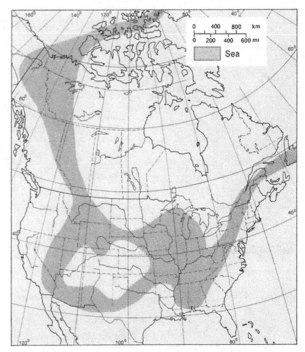

Figure 1.10 The shaded area is the approximate extent of the epicontinental sea in North America in Late Cambrian time. (Adapted from *Historical Geology* by Dunbar and Waage, 3rd ed., John Wiley & Sons, Inc.)

deeps. They seem to have been less than 300 feet (100 m) deep over most of their extent. Since the shallow waters admitted considerable sunlight and were well aerated, they and the sea floor teemed with life.

From Cambrian time on, such seas periodically flooded onto the continent and extensive, horizontal beds of marine sediments were deposited over parts of North America's older basement-rock platform in overlapping, irregular layers. Sometimes the marine sediments were deposited in one area, sometimes in another, for the seas advanced and withdrew, formed bays and straits, and shifted about as the earth's crust warped upward and downward and as sea level fluctuated. The sedimentary rock record left by the Paleozoic seas is more orderly and clear than that left by more ancient Precambrian seas.

The Paleozoic seas dominated the Wisconsin scene intermittently for something like 200 million years, from Late Cambrian through at least the Ordovician, Silurian and Devonian periods (see Plate 1).[2] Rock layers that formed during those period still coat parts of Wisconsin. Some or all of Wisconsin was probably

2. A geologic period is named for the place where rock of that age was first studied. The Early Paleozic systems were first identified in Great Britain in the eighteenth century. The Cambrian is named for Cambria, the old Roman name for Wales. Ordovician and Silurian are named for ancient British tribes, the Ordovices and the Silures, who inhabited the respective rock-type areas in Roman times. Devonian is named for Devonshire, England.

submerged even after Devonian time, but no sure geologic evidence remains; any rocks that may have formed then have since been eroded away. In Wisconsin the most continuous rock layer is the lowest one, Cambrian in age, protected longest by burial under younger rocks; and the least extensive layer is Devonian in age, uppermost and youngest, and most exposed to erosion.

The environment of the sea floor at any place was altered repeatedly because of changes in water depth and in the type and amount of material brought there. The turbulence of waves and the movement of currents varied in response to shoreline shape, water depth, and the wind's direction and velocity, and thus affected water clarity and stillness and the distribution of sediments. The intensity of sunlight penetrating the water increased and decreased. So in any location different sediments—coarse or fine, sand or mud or lime—accumulated at different times and are layered one atop the other. And as the environment changed in response to the water's depth, temperature, motion, and clarity, and to the composition and illumination of the sea floor, the kind of plant and animal life that could live there also changed. The makeup and arrangement of the resulting diversified beds of sandstone, shale, limestone, and conglomerate, and the fossils in them, help geologists to divide the rock beds into identifiable strata from youngest to oldest, and to reconstruct conditions existing at the time of their formation.

The Paleozoic seas in central North America were ever-changing. Because they were shallow, just a slight variation in depth could move a shoreline a considerable distance—hundreds of miles over low land. Of course, that move may have taken thousands of years. As the sea fluctuated in area and depth over Wisconsin, and entirely withdrew a number of times, the shoreline slowly migrated, relinquishing land at one time and reclaiming it at another. A given area would be under deeper or shallower water at different times, and sometimes above water, subject to erosion.

Some sedimentary rock layers are only a fraction of an inch thick, indicating that conditions changed after a relatively short time or that little sediment was available there. Other layers have great thicknesses of hundreds of feet, indicating a long time of steady conditions or an abundance of material brought into the water there. At some levels in the layering of the strata two horizontal beds that lie one atop the other may be separated by an *unconformity;* that is, an irregular surface or break in deposition where the younger and older beds do not merge smoothly, usually because a period of erosion or nondeposition intervened (fig. 1.11).

In chapters 4 and 5 we shall see how various Paleozoic sedimentary rock layers are distributed over the state, and how dissimilarities in their composition and resistance to erosion have helped determine the pattern of the landscape and regional differences in the land's productivity.

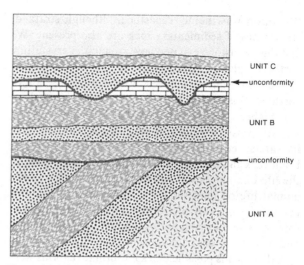

Figure 1.11 Examples of unconformities, where an erosion surface separates younger strata from older rocks. Rock unit A was eroded to a level surface before deposition of unit B. Later unit B was eroded irregularly before deposition of unit C.

The Cambrian Scene

Although Earth is judged to have been about 4 billion years old in Cambrian time, it was still biologically primitive compared to the modern world. Plant and animal life was confined to the sea. The land was without vegetation to slow the rain's flushing of rock debris toward the sea. Possibly pioneer lichen scales clung to some rocks like thin scabs; but more advanced land plants with roots, stems and leaves had not developed, or were so few and so fragile that fossil evidence has not been preserved.

Beaches in the Cambrian would have looked much like those of today, however, with waves planing off shorelines, cutting and building terraces, pounding during storms, or quietly rippling in. The waves, currents, and undertows rolled cobbles and gravels and sand grains over and over, just as they do now, rounding them, sorting them, and assigning them to locations where they would settle and stay for hundreds of millions of years.

The ancient shield was itself the source of most of the Cambrian sedimentary material being laid around its submerged edges. Much of the shield is granitic rock, which decomposes into quartz sand grains and finer particles. The fine clay and silt particles released by this and other decomposing rocks were dispersed by wind or carried by rivers into the sea, where they drifted in suspension away from shore. The more resistant quartz particles survived erosion's battering as sand grains. Larger and heavier than the fine particles, upon reaching the sea they stayed nearer the shore and accumulated in great quantities. Some of the shield was quartzite, and it too yielded quartz sand when it broke up. Therefore, the Cambrian rock that formed over

Wisconsin was mainly sandstone, although strata of other types of sedimentary rock are also present. We find this sandstone exposed now around the crystalline northern upland (plate 1). At one time sandstone left by the Cambrian sea covered most, and perhaps all, of northern Wisconsin, but it has since been eroded off that higher area, where it was relatively thin.

Wisconsin was near the equator in Cambrian time; its waters were tropically warm. Fossils found in Cambrian rock in and around Wisconsin tell something of the life of that period. Although there was no known animal life on land then, the waters held a complex community of early life forms, copious in number but small in size. No vertebrates (animals with backbones and bony skeletons) are known to have lived that long ago, though it is possible that some primitive forms were present (fig. 1.12). Cambrian marine plants were nearly all algae: some, just minute, single-celled types; some, seaweed; some, lime-trapping types that built stony, concentrically layered, round-topped stromatolites which, when occurring in aggregation, formed reefs (fig. 1.13).

One of the largest and most advanced animals of the time was the trilobite, named for its body's three-lobed appearance. Trilobites of many sizes and species crawled on the sea floor or swam near the bottom looking for food. Most were 2 to 3 inches (5 to 7 cm) long, but some were considerably larger. Trilobites reached their prime in Late Cambrian, dominating the world's animal realm in size and numbers. Thereafter they began a gradual decline toward extinction, which would come at the close of the Paleozoic (fig. 1.14).

Brachiopods were another common group of animals of the Late Cambrian sea. Having two shells, one above and one below, they looked somewhat like clams. Each of their shells was symmetrical—an identifying characteristic. They attached themselves to solid objects, and opened their shells to obtain food. These early brachiopods were small, and their shells were about the size and hardness of a large fingernail; later the shells of some became larger and harder. Brachiopods outlived the trilobites and, in fact, some survive today.

Spiral-shelled gastropods, early forms of snails, also inhabited the Cambrian sea, as did other shelled and soft-bodied animals. Tunnel holes and trails in rock, left while sediments were still soft, indicate that worms were present (fig. 1.15).

In Wisconsin no clear interruption can be found in the sequence of fossils or rocks to separate the Cambrian from the Ordovician, so here the boundary between these periods is somewhat arbitrarily defined. It is judged to have been about 475 million years ago.

Figure 1.12 Diorama of a Middle Cambrian sea floor in western North America. Among the feathery seaweed and primitive "organ-pipe" sponges are worms, trilobites (in foreground—two at extreme left and one just right of center), umbrella-shaped jellyfish, and other small animals. (Courtesy, Field Museum of Natural History, Chicago)

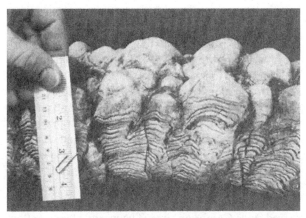

Figure 1.13 A close view of stromatolites shows their laminated, bulging structure. They built up on the sea floor, layer upon layer, by sediment collecting on sticky algal mats. Each day the algae grew upward and spread out forming a new mat, and sediment would collect on it. Stromatolites of much larger size formed reefs. These small ones are from Ordovician rocks near Blue Mounds, Wisconsin. (Geology Museum, University of Wisconsin-Madison. Photographer, Lawrence D. Lynch)

a

b

Figure 1.14 Trilobite fossils from southern Wisconsin. The pair are in Middle Silurian, Niagara dolomite. The single one is Middle Ordovician (in the Platteville Formation) from Fennimore, Grant County. Scale line for pair = 2 cm (0.8 in.); for single one = 1 cm. (Geology Museum, University of Wisconsin-Madison. Photographer, Lawrence D. Lynch)

Figure 1.15 Fossil "worm" burrows in Upper Cambrian sandstone near Ableman, Sauk County. Scale line = 1 cm (.39 in.). (Geology Museum, University of Wisconsin-Madison. Photographer, Lawrence D. Lynch)

The Ordovician Scene

In Ordovician time, as in Cambrian time, the Wisconsin region went through repeated cycles of submergence and emergence involving many fluctuations of water depth. During part of the Ordovician Period, North America was more widely flooded than at any time since, with half the continent inundated and the unsubmerged land separated into islands (fig. 1.16).

Although the eastern and western edges of the continent were experiencing strong geologic disturbances, structural changes in the older, stable Wisconsin area were gentle and slow. Ever since Cambrian time northern Wisconsin had been gradually rising or doming up. The overlying Paleozoic sediments therefore were also being uplifted, but so gently that they were hardly disturbed. They sloped upward slightly to the north as the old crystalline rock there rose under them, forming the *Wisconsin Dome*. They also warped upward in a low bulge known as the *Wisconsin Arch*, which trends north-south through the southcentral part of the state (see fig. 4.2 and plate 1). This arch is widest in its northern part, and narrows southward into Illinois. (It may be that the northern dome and the Wisconsin Arch, instead of rising, remained stable while areas alongside them sank, producing the same relative effect.)

Repeated fluctuations in the sea's depth and environment in Ordovician time left a variegated succession of sedimentary layers over the Wisconsin area, as in the Cambrian Period, but now sandstones decreased as shales and limestones increased. Just as the Cambrian rocks here are characteristically sandstone, so the Ordovician rocks are characteristically limestone, or, more specifically, dolomite. As we have already noted, most of the original limestone in the Wisconsin area

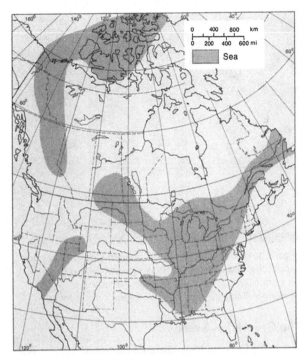

Figure 1.16 The shaded area is the approximate extent of the epicontinental sea in North America in Middle Ordovician time. (Adapted from *Historical Geology* by Dunbar and Waage, 3rd ed., John Wiley & Sons, Inc.)

was changed to dolomite by the addition of magnesium. During that gradual change recrystallization took place, which in some cases obliterated fossils and sedimentary structure. Dolomite is a more durable rock than pure limestone, and in this area it is more resistant and erodes more slowly than other sedimentary strata. Where a dolomite stratum emerges at the surface it commonly forms a ridge. One of the more evident and continuous Ordovician dolomite ridges is Military Ridge, which runs east-west across southwestern Wisconsin south of the lower Wisconsin River (plate 4).

The Ordovician sea, like the Cambrian sea, was shallow. It was clear and it was warm, subject to no severely cold season, judging by the plants and animals that lived in it. Wisconsin still was nearer the equator than it is now. The fossil record is lengthy and complex, but a few typical and significant representatives of it give a partial picture of Ordovician life in the Wisconsin area (fig. 1.17).

Advances and changes in underwater life had occurred since Cambrian time. The warm, clear, shallow waters were the required environment for corals, and they proliferated. The coral structures were built, as modern ones are, by small animals known as polyps, many of which lived in colonies. Each animal extracted calcium carbonate from the sea and secreted it as a limy substance to build a hard, protective wall around itself alongside its neighbors. Only the top of its body appeared over the edge of its "apartment." The innumerable walled housings of these tiny animals, combined in communal structures, multiplied as the polyp population increased. After the occupants were gone the vacated housings remained, like porous rock. In time that hard, limy material was incorporated in the rock-forming deposits of the sea floor.

Bryozoans, sometimes called "moss animals," are another of the various lime-secreting organisms that were present in the Ordovician sea. They survive today. They live in colonies that look like plants, and the skel-

Figure 1.17 Diorama of a Late Ordovician sea floor depicting life of the period, including long, tapered cephalopods, seaweed, spherical coral colonies, antler-like coral bryozoans, brachiopods, clams, snails, and trilobites. (Courtesy, Field Museum of Natural History, Chicago)

14

etal structures that some of them build are of a delicate, branching shape.

Trilobites, though still numerous, had lost their leading status to the symmetrically shelled brachiopods. Shelled animals abounded in a multitude of varieties, many resembling modern clams and snails. An interesting new type was the cephalopod, an octopus-like creature that consisted of a head surrounded by tentacles, housed in a chambered shell which was either coiled or straight and tapering. Some cephalopods crawled, some swam, and all were predators. They were the largest animals in Ordovician time. The straight shells of some were more than 15 feet (5 m) long.

The crinoid, an animal that resembles a plant, appeared in the Ordovician sea. It is sometimes referred to as "the sea lily." From where it "roots" itself to the sea floor a tall, thin "stem" made of hard disks extends upward, and at the top is its main body—round, with radiating arms, having the appearance of a flower. The stem moves gracefully in the water, and the body floats and feeds on passing small sea life.

During the Ordovician simple vegetation probably had taken hold on land, but generally the above-water scene must have been bleak.

The Silurian Scene

Whereas the Cambrian and Ordovician systems of rocks in Wisconsin are quite complex, with numerous alternations of strata, the Silurian which follows is more homogeneous. It is almost all dolomite. We infer from this that during the Silurian Period the areas of marine deposition in Wisconsin contained an abundance of lime-secreting life, and were receiving very little muddy or sandy sediment. Either eroding land was farther away, or by then the land was low and erosion had slowed considerably.

The Silurian began an estimated 425 million years ago and lasted until about 413 million years ago. In Early Silurian the sea encroached upon the continent in several embayments. One from the northeast penetrated into Wisconsin and covered at least the eastern and southern parts of the state, where Silurian rocks are found. Even more of the state was inundated during subsequent Silurian time (fig. 1.18 and plate 1).

Again, this sea was shallow, warm and clear, an ideal habitat for coral. And now corals thrived in dense colonies and built extensive complexes of reefs in southern and eastern Wisconsin, as well as in areas beyond and in other seas of the world. So prolifically did these organisms build that the Silurian is known as the Age of Coral. Coral reefs were the "cities" of the Silurian sea, the centers of activity, heavily populated with numerous varieties of animals and luxuriant with plant life (fig. 1.19).

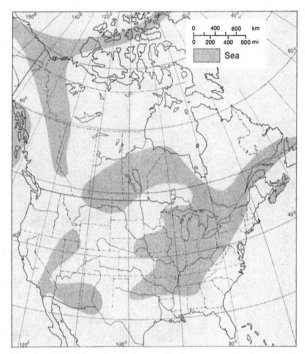

Figure 1.18 The shaded area is the approximate extent of the epicontinental sea in North America in Middle Silurian time. (Adapted from *Historical Geology* by Dunbar and Waage, 3rd ed., John Wiley & Sons, Inc.)

The marine plants and animals were carry-overs from earlier Paleozoic seas, but now many were more diversified and had evolved into new forms, while others were decreasing in number in the face of competing types more fit for this environment. More than forty species of coral were constructing their compartmented homes in intricate designs of honeycombs, chains, horns, stars, cups, and organ pipes (fig. 1.20). Bryozoans were also building their stony structures, delicate and branching, but in fewer numbers than before. Tall-stemmed, flower-like crinoids swayed gracefully upon the reefs, in some places in profusion. Small shelled animals crept over and around the reefs, or attached themselves to the reefs or sea floor and waited for their food to swim or drift by. A variety of trilobites crawled or swam about. Brachiopods of many species and sizes continued to compete successfully, and in some places crowded together in thick clusters like oyster beds, clinging to solid objects. Other shelled animals, including gastropods, added to the scene. Primitive fish were present, but scarce. Large octopus-like cephalopods were there, preying on other animals; they had been the supreme animal of the sea, but now sea scorpions had apparently become the largest animal in the world (fig. 1.21). And there were millipedes—elongated crawling animals with numerous legs. They and the scorpions may have started to become air-breathing at about this time, and may have been among the first

Figure 1.19 Diorama of a section of coral reef such as might have existed in Wisconsin in Middle Silurian time. Coral colonies are in the center and to the left. Crinoids wave like flowers in the water. Trilobites, snail-shaped cephalopods, clams, and brachiopods also inhabit the area. (Courtesy, Field Museum of Natural History, Chicago)

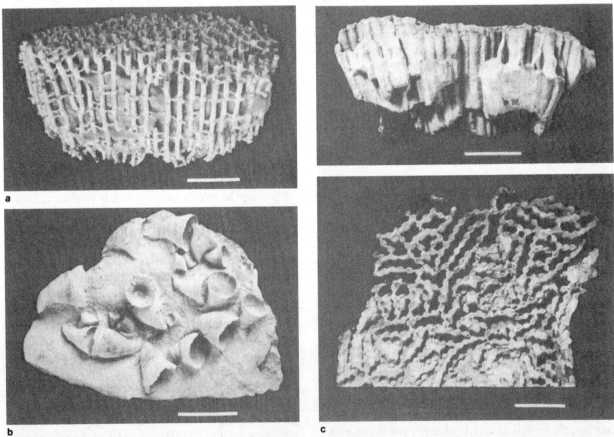

Figure 1.20 Specimens of coral, which used to house tiny animals. Scale line = 2 cm (0.8 in.). **(a)** "Organ Pipe Coral" (*Syringopora*) from Silurian rocks in Door County. The colony is held together by tiny crossbars. **(b)** Solitary "Horn Corals" of the genus *Strepelasma* from Middle Ordovician rocks at Beloit. **(c)** *Halysites,* the chain coral, from Silurian rocks of Door County. Side and top views. (Geology Museum, University of Wisconsin–Madison. Photographer, Lawrence D. Lynch)

Figure 1.21 Sea scorpions (Eurypterids) were perhaps the largest animals of the Silurian. In this diorama they are shown with snails, crustaceans and worms among the seaweed. Some relatives of these sea scorpions may have started to become air-breathing by this time, venturing onto land. (Courtesy, Field Museum of Natural History, Chicago)

animals to venture onto land. In Late Silurian time simple vegetation was appearing on land, probably mainly near margins of water bodies. No fossils of Silurian land plants have been found in Wisconsin.

The Silurian sea floor slowly sank, but material accumulated on it rapidly enough to maintain the shallowness of the water. For an estimated 12 million years sea-floor materials collected and compacted to form rock strata with a total thickness of about 600 feet (180 m) in Wisconsin, (plate 5).

The coral reefs themselves supplied much of this rock-making material. The corals were always building toward the surface, adding to their tops, while their bases were being buried by the sea's waste on the ever-thickening ocean floor. Fragments of broken coral fell around the reefs, and multitudes of whole and broken shells of marine animals dropped there too. This calcareous debris was agitated in the water and shattered. In, on, and around the reefs accumulated the broken and decayed remains of everything that had lived upon the reefs and in their many nooks and hollows, and the incessant settling "dust" of myriads of dead microscopic organisms, which have been continuously sinking down through ocean waters ever since sea life began. Spaces in the reefs were filled with fine material; calcareous sand spread out on the sea floor; and away from the reefs the floor was a white, limy mud.

The Silurian dolomite which formed from the sea-floor materials is the most resistant Paleozoic rock in the state. Generally where an edge of this stratum is exposed it forms a ridge or low cliff. And the strata it covers are protected because its presence slows erosion. Its protective cover is what has permitted Door Peninsula to exist while softer rocks east and west of the peninsula have been removed. Its high western edge follows and overlooks a north-south lowland of softer rocks occupied by Green Bay, the Fox River, Lake Winnebago, Horicon Marsh, and the Rock River. A remnant of this dolomite forms a cap on Blue Mound in Iowa County. This and several other remnants of Silurian dolomite were left as detached islands, or *outliers,* when the main rock layer was dissected and worn back into Illinois and Iowa.

All that now remains of the Silurian rocks in Wisconsin is a solid north-south strip down the eastern part of the state from the tip of Door Peninsula into Illinois, and isolated caps on a few outlier hills in southwestern Wisconsin. The strip along Lake Michigan is about 40 miles from west to east in its widest places, which are south of Fond du Lac and at the southern boundary of the state. It is slowly being eroded eastward toward Lake Michigan, where it slants down under the lake and extends on to the east. This lakeshore strip is largely covered by glacial deposits, most thickly in the southern part, but its western edge outcrops noticeably in many places.

The eroding edge of the strong Silurian dolomite can be traced an unusually long distance—from eastern Wisconsin north through Door Peninsula and Washington and Rock islands, through the peninsulas and islands separating northeastern Lake Huron from Georgian Bay, across Ontario to the state of New York where it forms the escarpment that creates Niagara Falls (see fig. 5.2). The area between Niagara Falls and Wisconsin was part of the same arm of the Silurian sea, and the rock throughout is similar. This especially resistant rock is known as "Niagara dolomite" even in Wisconsin. In northeastern Illinois it is completely buried, but after curving westward across northern Illinois its eroding edge surfaces and is prominent again in northwestern Illinois and eastern Iowa.

The Devonian Scene

A break in marine deposition occurred in Wisconsin after the Silurian, so Early Devonian rocks are absent. Deposition took place again in the Middle and Late Devonian, from roughly 400 to 350 million years ago. The resulting rocks consist mainly of dolomite, with thin layers of shale interspersed and a layer of shale on top. It is clear that the Devonian sea submerged part of southeastern Wisconsin, for its rocks are found along the Lake Michigan shore. How much more of the state that sea covered is uncertain, for no known Devonian outliers remain west of the shoreline strip. However, the Devonian sea did cover a large area to the east; its strata slope eastward under Lake Michigan, as do the Silurian strata below them (fig. 1.22).

In Early Devonian, land vegetation evolved into large, more complex plants, but the landscape was still primitive in aspect. Most of Wisconsin apparently remained above the sea, and it was probably greening with vegetation, but not the kind we know. Trees, flowers and grasses had not yet evolved. Plants were low and without leaves. The largest known plants of Early Devonian were perhaps a foot or so high, consisting only of branches with spore cases at their tips; some bore the rudiments of leaves. By this time the earliest small air-breathing animals had arrived.

In Middle Devonian, on swampy shores along the inland sea, stood forests of scale trees, so called because their leaves were close-set and left scars that looked like scales on the trunk and limbs. Ferns, rushes and early conifers also decorated the land, and perhaps were present in Wisconsin. A habitat hospitable to land vertebrate animals was developing.

In Late Devonian vegetation spread over more of the land, and plants advanced further and grew larger.

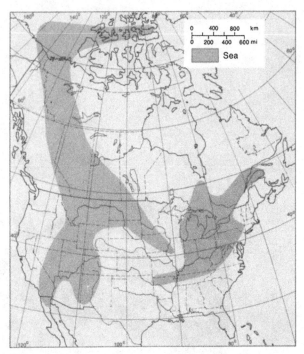

Figure 1.22 The shaded area is the approximate extent of the epicontinental sea in North America in Middle Devonian time. (Adapted from *Historical Geology* by Dunbar and Waage, 3rd ed., John Wiley & Sons, Inc.)

Through the following Mississippian and Pennsylvanian periods they would attain unprecedented denseness and tropical luxuriance. Their abundant remains, which accumulated in swamps, would form coal, but none is found in Wisconsin (except a few thin seams). If coal-bearing rocks ever existed in Wisconsin's terrain they have been eroded away. (Such rocks are found as close as Illinois and Michigan.)

In the underwater environment evolutionary progress had taken place between the sea's Silurian withdrawal and its Devonian return. Primitive fish of many kinds were in the seascape; they became so important that the Devonian is known as the Age of Fishes. Vertebrates and air-breathers were developing, leading the way for higher forms of life to come. Some fish could breathe air part of the time, and there were early amphibians, which combined traits of fish and reptiles and spent part of their life on land. Waters of Wisconsin's Devonian sea floor may have been somewhat muddy, not as transparent as before, as corals and crinoids were no longer abundant here. Giant cephalopods had gone elsewhere. But the sea swarmed with other shelled animals, including the enduring, diversifying brachiopods. The gastropod population had lessened, and trilobites were fewer than in the past.

The Layering of Sedimentary Rocks

So it was that from Late Cambrian to Devonian time sedimentary strata built up, one atop the other, to a great thickness (plates 1 and 5). As we look at roadcuts, cliffs, and quarry walls in sedimentary rock we can see the layering and notice how one type of rock gives way to another, and how the strata differ from one location to another. Cores taken from well drillings reveal the layering too. Because most of the rock laid down by the ancient seas is buried and invisible or has been removed by erosion, it is impossible to construct a complete, connected account of what happened long ago. Even where clearly exposed, the bedrock layers are not all distinct. Many are thin or taper out over a short distance, some have been reworked by water during a later submergence or dissected by streams during emergence, and confusing strata sequences occur from place to place. So it is no simple task to identify strata and trace them through a broad area. Still, certain characteristics of the deposition process help in learning the organization of these rock'pages of Earth's diary.

Conglomerate was often the first in a series of strata laid down as a sea moved inland. At the shoreline, breakers tore pieces of rock from land standing in their path. These and rock fragments lying on the old land surface were rolled by waves and rounded and left on the sea floor as the breakers advanced inland beyond them. This mixture of rounded rocks of various sizes then would be buried by a more homogeneous layer composed of sand, which normally was deposited next as the wedge of water proceeded up the sloping shore.

And as the inward-moving waters deepened over that spot, a muddy layer of fine silt and still finer clay (which would become shale) might be laid over the sand (which would become sandstone).

As one looks at a cross-section of sedimentary rocks, as in a cliff, and sees a layer of conglomerate rock merging upward into sandstone, and the sandstone merging into shale above it, one can visualize in retrospect an invading, rising sea. Its floor was first beach rubble that its waves helped to make, then sand, then silt and clay, all in that same location, one over the other. Then if the next layer on top reverts from shale to sandstone, one envisions the sea retreating. Or if instead the shale merges upward into limestone or dolomite, one could deduce that the sea had spread farther inland, that water had become deeper there, or that for some other reason silts and clays had failed to reach that site. (More will be said of this depositional sequence and its variations in chapter 4.)

Of course, if there is a break in the sequence of rock strata, where erosion had worn off some sedimentary layers before deposition resumed, one may not know how many layers were removed from the sequence. To learn how long the erosional interlude lasted, one looks at the fossils in the individual rock layers. Fossils from the interrupted sequence in question will be compared with fossils in strata elsewhere, where no interruption occurred, to determine what is missing. This is one way geologists piece together the past, reading rock profiles wherever they can and matching up the findings from many sites (fig. 1.23).

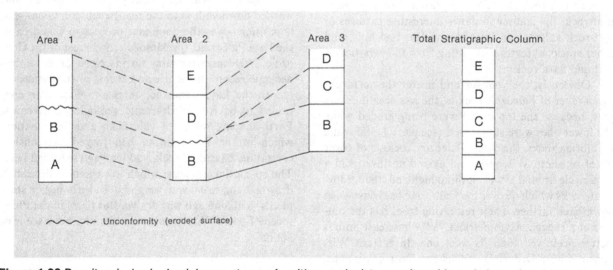

Figure 1.23 Despite missing bedrock layers at unconformities, geologists are often able to deduce what the total stratigraphic column of an area was by matching known sedimentary strata of different places.

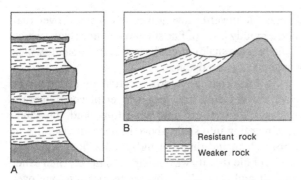

Figure 1.24 Examples of differential erosion. Weaker rock erodes more easily and faster than more resistant rock.

Resistant rock

Weaker rock

The Pattern of Paleozoic Strata in Wisconsin

As we survey the arrangement of Paleozoic strata in Wisconsin, a pattern becomes apparent. The strata have been generally eroding away from the northern dome and the Wisconsin Arch, toward the east, south and west, retreating in the direction that the strata dip. The rock beds were beveled as erosion wore off the top of the high northern area and, to a lesser degree, the southcentral area, and as a result beds of all periods are exposed. As a simplified illustration, imagine a pack of many colored sheets of paper (representing sedimentary strata) bent up slightly through the center, with the raised part sliced off horizontally. Then all of the colored sheets are in view as bands of color.

Some of the exposed Paleozoic strata were resistant to erosion, while others eroded more rapidly (fig. 1.24). Because of this *differential erosion*, which removes weak rock at a different, faster rate than resistant rock, the landscape shows interesting patterns of soft-rock valleys and hard-rock ridges and hills, and other erosional features resulting from the variation of soft and hard rock.

Obviously, the younger and nearer the surface a given layer of Paleozoic rock is, the less area it covers now, because the top layers were being eroded while the lower ones were still buried (see plate 1). So while Cambrian rocks, the oldest Paleozoic rocks, still cover all of southern Wisconsin and extend northward in a semicircle around the crystalline highland, the Ordovician rocks which lie over the Cambrian ones have been worn back farther. Their retreating front has the outline of a ragged, asymmetrical "V" whose east limb is better preserved than its west one. In eastern Wisconsin where Ordovician rocks are most continuous, they are still in large part overlain by younger rock. In southern Wisconsin they are being dissected; and in western Wisconsin, where dissection has progressed still further, they are divided into separate patches or outlier hills.

Silurian rocks, still younger and higher in the sequence, cover even less area than the Ordovician—just the strip along Lake Michigan and some outliers. And the youngest Paleozoic rocks, those of the Devonian Period, cover the least area of all: a meager fringe along the Lake Michigan shore from Sheboygan to Milwaukee.

The Long Unknown Interval

After the Devonian Period ocean embayments repeatedly spread onto and across the continent, plastering layer upon layer of sediments on and around the old platform base; but gradually the fluctuating ocean withdrew from the lands it had built, to its present position. It is not known whether the sea invaded Wisconsin again after the Devonian. Very likely it did so in the Mississippian Period which followed; Mississippian rocks are seen as nearby as the Michigan shore of Lake Michigan. The sea may have returned even after that, leaving sediments that have since been eroded. Or, on the other hand, Wisconsin may have stayed above the sea in the remaining Late Paleozoic, as well as in following eras—the Mesozoic (the Age of Reptiles, including dinosaurs), and the Cenozoic (The Age of Mammals, when humans appeared). (See fig. 1.9.)

In the Mesozoic Era the Atlantic Ocean was opening up. While North America and Europe slowly separated, while they and other continents "drifted," and while many affected areas were deformed into mountains, Wisconsin at North America's core reposed undisturbed except for gentle warping. Its surface material was continually being removed and washed down valleys to the sea. Because erosion—not deposition—was the dominant process in this area in the Late Paleozoic, the Mesozoic, and most of the Cenozoic, Wisconsin contains no inscribed or entombed geologic record of the life and events of those times.

In the Late Cenozoic, quiescent Wisconsin continued to be free of dramatic geologic occurrences. Earth-shaping forces of volcanism and deformation, which in the Precambrian had forged Wisconsin's crystalline basement rock, had long ago subsided here. The epicontinental seas, which in various outlines had drowned the area and weighted it with heavy sediments, had long ago withdrawn. But then, in the Pleistocene Epoch, Wisconsin once more became a scene of action.

The Pleistocene Ice Age

In the Pleistocene Epoch Wisconsin was invaded again, not by water this time, but by a more powerful agent—ice. The glacier was deeper and heavier than the epicontinental seas had been, and unlike the seas, was frigid and opaque. What land it covered became lifeless. There was only the grinding of rocks and grit

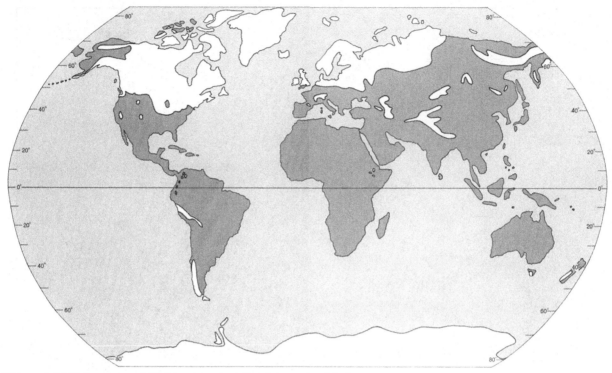

Figure 1.25 Major areas of Pleistocene glaciation (shown in white).

in the darkness under the smothering ice as it inched its way along, and the washing out of ground-up rock debris whenever melting occurred.

This was the Ice Age when huge, thick ice sheets covered northeastern North America, Greenland, Antarctica, northwestern Europe, and parts of Siberia, and smaller ice caps and valley glaciers frosted mountain areas around the world. However, most of the rest of the world away from the ice remained normally warm (fig. 1.25).

Although "the Pleistocene" and "the Ice Age" are often used interchangeably, the terms are not synonymous, nor were those times synchronous. "The Pleistocene Epoch" is a designated geologic-biologic time period that has been assigned a beginning point of about 2 or 3 million years ago. "The Ice Age"—*the* Ice Age— is a popular term signifying that time preceding the warm Present when glaciers grew immense and overran nearly a third of the earth's land surface. (That term does not apply to pre-Pleistocene glacial periods, though any time of large-scale continental glaciation may be called an "ice age.") Some earth scientists would have the Ice Age begin when small, scattered glaciers were forming and expanding somewhere in high latitudes and mountains far from Wisconsin, back in the earlier Cenozoic (the Tertiary Period) many millions of years ago. Some would not have the Ice Age start until much later when broad, continental ice sheets were taking shape. Others have still different opinions about when the Ice Age and the Pleistocene began, based not only on glacial or geologic events, but on climatic and biol-

ogic changes throughout the world. ("Pleistocene" means "most recent" in reference to animal life.)

Because the Pleistocene is the geologic epoch closest to the Recent, in which we live, it would seem that its events should be clearly understood. Yet some of the most elementary questions about it remain unanswered. We do not know when the glaciers began to grow—or where—or why. We do not know why great ice sheets spread out as they did, then melted away, then returned, going through this cycle several times. Nor can we predict whether the ice will return sometime in the future.

By the time the Ice Age began, the continents were in approximately their present locations and Wisconsin was geographically positioned where it is now. The ice sheets that entered Wisconsin time after time apparently had their origin in eastern Canada in the vicinity of Hudson Bay. They must have begun forming under conditions of reduced temperature or increased snowfall, when more snow fell in Canada in the winters than could melt away in the summers; so over many years leftover snows accumulated until the compacted snow turned to ice and became so thick—probably about 200 feet (60 m)—that it began to move under pressure of its own weight. As an ice sheet grew thicker over northeastern Canada it bulged outward in all directions— toward the Arctic Ocean, toward Greenland (which had acquired its own ice sheet), and westward to near the Canadian Rocky Mountains, which sent out smaller glaciers to meet it. But an ice sheet spread most vigorously toward the south, because its main moisture

Figure 1.26 Pleistocene glaciation in North America.

supply for snowfall came in humid, warm air masses from the Gulf of Mexico and the Atlantic Ocean. The ice sheet thickened to become a mile high, and possibly, it has been estimated, as much as 2 miles (3 km) high in places. It spread like pie-crust dough being rolled out; its advancing front had a scalloped outline, and the more rapidly moving sections pressed ahead in separate lobes. Each ice sheet that came down from the north had a somewhat different shape, and covered a somewhat different area, but considering all ice sheets together they reached as far south as (roughly) the Missouri and Ohio rivers in the central United States, and New York City in the east. The ice pushed out into the Atlantic Ocean onto the continental shelf. Sea level around the world dropped because of the great amount of moisture held in the world's glaciers. The mountains of western Canada and southern Alaska together developed their own local ice sheet, much smaller than eastern Canada's; and lesser glaciers formed in many mountains of western United States (fig. 1.26).

Like today's glaciers, the ice moved ponderously slowly, sometimes only a few inches or feet a year. At times it stood still for long periods, or temporarily melted back a way. At other times it surged forward so "fast" that its movement would have been perceptible to an observer.

As the giant ice sheets from Canada pressed slowly into Wisconsin and beyond, each in turn overrode the highest hills in its path, knocked down forests, erased rivers, and scraped and ground up the surface it passed over. After occupying the land for thousands of years the ice would melt back, pouring out floods of meltwater, dumping its rock waste over the terrain, and leaving the barren land to restore itself, which it always did. The ice left its mark on Wisconsin so vividly and in so many ways that this state is renowned as a repository of glacial features and is a textbook of glacial history.

Material deposited by the ice and its meltwater is known as drift. Drift may consist of material of any size, from fine particles to boulders, sorted or mixed. About four-fifths of the state was covered with drift or with temporary glacial lakes that the ice had created by damming drainageways. Southwestern Wisconsin and the northwestern corner of Illinois are not drift-covered and do not show the signs of having been overrun by the ice sheets, as all the surrounding area does. That *Driftless Area* is a window into the past where one can see how the rest of southern Wisconsin and much of the adjacent area might look now if the ice had not come and altered the landscape (Plate 2).

The Ice Age was not all cold. There were interglacial periods when the ice sheets temporarily disappeared, and the climate in areas that had been glaciated was mild for thousands of years—sometimes even warmer than it is now. The bare, rock-strewn land became softened with new coverings of soil and vegetation. No evidence has been found of human habitation in the Americas early in the Ice Age, but by the time of the last ice sheet Paleo-Indians (nomadic hunting people who predate modern Indians) were already on the continent. Presumably all or most of their predecessors came from eastern Asia by way of Alaska and moved southward even while ice covered much of northern North America.

When the last ice sheet retreated north into Canada the terrain had its present general configuration. Rivers flowed in their present courses; the Great Lakes were a new feature on the landscape; and smaller lakes dotted the glaciated surface. Plants soon reappeared on the ice-free land, animals returned, and with them into Wisconsin came the Paleo-Indians.

The first vegetation to clothe the cool, damp land must have been like that of the tundra in subpolar regions today—lichens, mosses, shrubs, and low, leafy and flowering plants; but there were no trees. This vegetation spread northward as the ice retreated, and so did the animals that grazed on it, including the caribou and musk-ox which survive in northern North America, and the now-extinct woolly mammoth, which was the size of the modern elephant and had thick shaggy hair, humped shoulders, and immense, long, curving tusks (fig. 1.27).

Next the hardy, moisture-tolerant spruce, tamarack and willow trees appeared. In that boreal forest environment browsed the mastodon (fig. 1.28). It was stockier and somewhat shorter in height than the mammoth, and its tusks were smaller and only slightly curved. In swamps lived the giant beaver, more than 7 feet (2 m) long and weighing up to 500 pounds (225 kg). Deer, wood buffalo, elk, bear, and the forest pig were also present, as well as most of the smaller animals found in Wisconsin today (fig. 1.29).

With an increasingly warmer climate came pine trees, and later the mixed evergreen-deciduous forests. By 8,000 years ago the climate had become warm and dry enough for prairie grass to replace forest in much of southern Wisconsin. The scene began to resemble that of today.

Figure 1.27 Herds of shaggy woolly mammoths roamed the tundras and steppes of Ice Age Eurasia and North America. They probably used their long, curved tusks to clear snow away from the low vegetation, which was their food. (By permission of artist Zdenek Burian. ARTIA Foreign Trade Corporation, Prague, Czechoslovakia)

Figure 1.28 Much of Wisconsin looked like this late in the Ice Age. The ice had withdrawn and the climate had warmed enough to permit the growth of spruce forest and some deciduous trees. Here mastodons browse over poorly drained land as geese and other birds fly overhead. (Illinois State Museum, Springfield; Robert G. Larson, artist)

A Present-Day Perspective

Wisconsin's physical landscape is still changing. Not only are the natural agents of erosion and deposition slowly resculpturing it as they have always done, but now also the human hand is adding its touch as it grades, plows and paves the surface; mines and quarries the rock; digs out glacial sands and gravels; redirects streams; drains or creates marshes and lakes; changes shorelines; and alters vegetation, a controller of erosion.

As we travel about and see exposures of solid bedrock and the covering of loose rock, including glacial drift, we are keenly interested in these earth materials—their composition, distribution and arrangement—not only because of what they reveal about landforming events of the past, but because of their roles in determining the value and use of the land today, and of the effect they have, directly and indirectly, upon human lives. They are a basic factor in the way an area develops, both naturally and culturally. They determine the levelness or ruggedness of the terrain and influence its drainage system of rivers, lakes and marshes. Porous rocks hold more groundwater than compact ones, and certain rocks can convert to more fertile soils than others. Some earth materials contain valuable minerals, or are useful in themselves for construction work or land improvement. The esthetic appearance of geologic features makes us enjoy seeing or being in certain areas more than others. In many ways the geologic makeup of a region helps to determine where people live, where farms prosper, where industries develop, where transportation routes lie, where tourists go, and it even influences regional architecture, for local materials figure in much of the building that is done.

Though Wisconsin's bedrock structure may appear plain when viewed superficially, smoothed and heavily masked as it is by weathered material, glacial drift and vegetation, we know it is anything but plain. Even as limited a look as we have had thus far tells us that every part of the state displays a tangible record of the events associated with its formation and its changes through time. Interested observers who have learned the language of geology can read this record. They can also look into the future and see how natural and human forces may further remodel Wisconsin's geologic landscape, and how its resources may best be utilized.

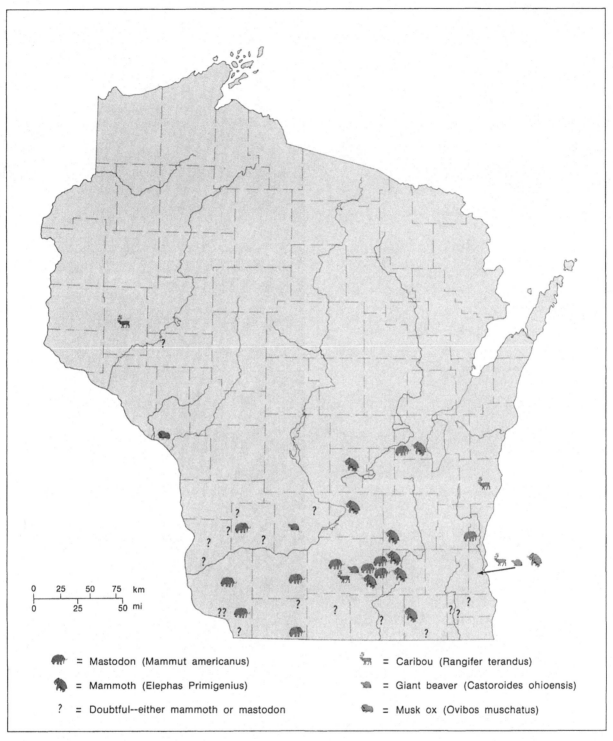

0   25   50   75   km
0   25   50   mi

🐘 = Mastodon (Mammut americanus)

🦬 = Mammoth (Elephas Primigenius)

? = Doubtful--either mammoth or mastodon

🦌 = Caribou (Rangifer terandus)

🦫 = Giant beaver (Castoroides ohioensis)

🐃 = Musk ox (Ovibos muschatus)

Figure 1.29 Authenticated sites where remains of certain post-Pleistocene mammals have been found. (John E. Dallman, including data taken from O. P. Hay and A. W. Schorger)

25

2

The Cryptic Crystalline Shield

Traveling onto Wisconsin's Precambrian shield area from the sedimentary rock area skirting it, one is hardly aware of the increasing elevation because the rise is imperceptibly gradual. From south to north across the central part of the shield, elevation increases less than 6 feet (2 m) per mile. But one might observe that the rivers flow outward from northcentral Wisconsin in all directions—toward Lake Superior, Green Bay, and the Mississippi River. The state's central and main internal river, the Wisconsin, has its source on the northern dome in Lac Vieux Desert (pronounced Lahk Vyoo Day-zair) on the state border in Vilas County.

The shield area has an elevation of about 1,400 to 1,650 feet (425 to 500 m) above sea level. Most of it displays a gently rolling or undulating surface. Divides separating drainage basins are low, and only here and there does one encounter high hills, ridges, escarpments, or deep valleys that interrupt the levelness (fig. 2.1). The shield area has been glaciated and covered with drift (plates 1 and 2). Most surface irregularities are caused by the uneven thickness of the drift, and many depressions in the drift contain lakes, swamps or marshes. The highest point in the state—Timm's Hill,

Figure 2.1 Because of long erosion much of the Precambrian shield surface is fairly level with wide valleys and low interstream areas, as in this scene near Wausau where the glacial drift cover is thin. (Robert W. Finley)

elevation 1,952 feet (595 m), in southeastern Price County—is a mound of glacial drift atop the bedrock shield (fig. 6.25).

While glaciation is partly responsible for creating irregularities in the topography, it is partly responsible also for smoothing the surface, because the ice ground down high features that it passed over, filled valleys with drift, and buried most of the old Precambrian hills. Generally the terrain's *relief* (the local difference in elevation between the highest and lowest points) is low, averaging no more than about 200 feet (60 m) in most areas.

Although the cover of glacial drift, the "skin" over the shield, can be broadly described without much difficulty, the shield's bedrock, its "body," is complicated in the extreme, and it alone is the subject of this chapter.

Lawrence Martin, in his well-known book *The Physical Geography of Wisconsin,* referred to the Precambrian shield as "The Lost Mountains of Wisconsin" and likened its remaining, contorted bedrock to stumps and roots of trees that are now gone,[1] for there are signs that mountains may once have stood in parts of the shield in Wisconsin. There was much tilting and folding of rock, much faulting (fig. 2.2), and numerous intrusions and extrusions of molten rock (fig. 2.3).

If erosion was rapid, rising landforms may not have reached mountainous proportions. They could have been worn down steadily even while they were being created by upthrusts and volcanic upwellings, and may have reached only moderate heights, just as much-used fingernails may stay short though they are growing.

The size and growth pattern of mineral crystals in a body of igneous rock indicate how slowly the magma cooled and how far below the surface it may have crystallized into rock. The more slowly the magma cooled,

1. Lawrence Martin, *The Physical Geography of Wisconsin,* Wisconsin Geological and Natural History Survey Bulletin No. 36 (Madison: The State of Wisconsin, 1916), p. 347.

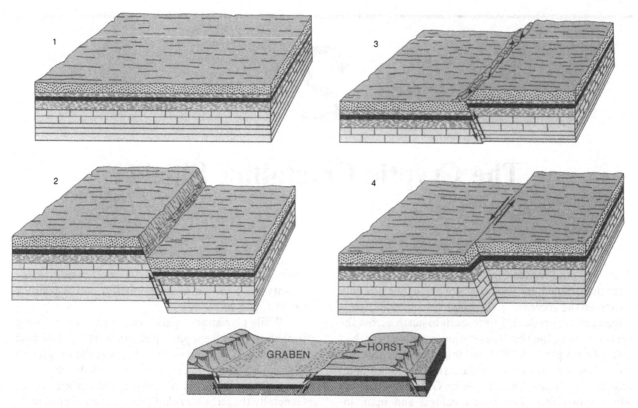

Figure 2.2 A *fault* is a fracture or fracture zone in rock along which there has been slippage and displacement of the sides relative to one another, in any direction. Some types of faults are illustrated here.

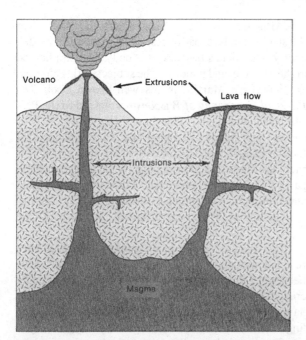

Figure 2.3 Magma within the earth, hot and fluid, sometimes, *in*trudes the hard crust above, working its way up through cracks and weak zones in the bedrock. If it reaches the surface it is then termed "lava" and *ex*trudes either in a gentle horizontal flow or as a violent volcanic eruption.

the larger the crystals of the resulting rock could be, for there was more time for like molecules to organize and for crystals to grow. Much igneous rock now at the surface contains crystals of a size that indicates they may have formed miles underground.

The Complexity of the Precambrian Rocks

Of all the geologic regions of Wisconsin, the crystalline shield has the most complex history. Most of it is unseen and unknown, masked as it is by overlying drift which in places is several hundred feet thick. Its rocks have gone through so much distortion and recrystallization that they read like a mixed-up cryptogram. Those parts of the shield that are exposed or that showed mining potential were the first to be geologically explored and mapped in detail; but over most of the shield there was only generalized or spotty mapping of bedrock. Recently more advanced and detailed mapping has been done in some areas.

Small-scale geologic maps have commonly utilized the simplest, least confusing way of depicting the heterogeneity of Precambrian rocks by showing all or a large part of the shield in a uniform pattern or color, and explaining in the legend that the seemingly homogeneous region is really composed of "undifferentiated crystalline rocks" that are "not surveyed in detail," or some other convenient generalization to that effect. Actually, on small-scale maps that is all a cartographer can do without hopelessly confusing the map-reader, or making an illegible map, because of the region's complexity.

The shield's many varieties of rock cannot be described without highly technical discussions of rock composition, which limitations of this book do not permit. A few significant types will be described now, and others later when description is pertinent.

Most of Wisconsin's Precambrian rocks were igneous in origin. A high proportion are metamorphic; that is, altered igneous and sedimentary rocks. Only in the northwest corner of the state (the Keweenawan area) does the shield contain unmetamorphosed sedimentary rocks.

Granite is found in widespread abundance in the shield of Wisconsin. Most of it was formed by the solidifying of molten material, but some may have been formed by the extreme metamorphism of certain preexisting rocks. It is found in a range of colors—gray, pink, red; and of textures—from fine-grained to coarse-grained. "Ruby red granite" is the official state rock.[2]

Locally occurring rocks that are related to granite are diorite, monzonite, and syenite (which are coarse-grained and have a mineral composition close to that of granite), and rhyolite (which has the same minerals as granite, but is finer-grained). Other common igneous rocks found in Wisconsin are basalt, a fine-grained, dark, solidified lava; and its intrusive counterpart, gabbro, a coarse-grained, dark rock.

Anorthosite is a coarse-grained rock related to gabbro in origin, but (being mainly plagioclase feldspar) it is lighter in color. Amphibolite is a dark, heavy metamorphic rock, sometimes greenish black, with parallel arrangement of mineral particles; long, slim hornblende crystals are common in its composition.

"Greenstone" is a general term given to metamorphosed igneous rocks that have a greenish color resulting from the presence of certain minerals. Parts of the shield in Wisconsin, neighboring states, and Canada have been termed "greenstone belts" because in them greenstone is the most commonly exposed rock. Many of these belts have been found to contain valuable ore deposits. Exploration has shown that other rocks associated with greenstone belts, including metamorphosed sedimentary rocks, may also contain sought-for mineral deposits, and so the term "greenstone belt" is often replaced by "volcanic-sedimentary belt."

Also found in many parts of Precambrian Wisconsin are various kinds of schist (which is foliated and can easily be split into flakes or slabs) and of gneiss (which is partly foliated and often has light and dark parallel layers). (See fig. 1.7.) Both may be either metamorphosed igneous or metamorphosed sedimentary rock. Of the metamorphosed sedimentary rocks, slate and schist are common because shale, from which they were derived, formed widely in Precambrian time. Argillite is hard, compact metamorphosed mudstone

which resembles slate, except that it does not break readily into thin plates as slate does. Quartzite, which is metamorphosed sandstone, is generously present, since much sandstone formed in Wisconsin in Precambrian time.

The record imprinted in the rocks, of what Precambrian events occurred, and in what time sequence, is only partly deciphered and decipherable. It is harder to understand the formative history and comparative ages of the shield's intermixed Precambrian rocks than of the younger Paleozoic sedimentary strata found over the rest of the state, which are relatively undisturbed. Because there was so much folding, faulting and volcanism in the shield, along with erosion and deposition, various types of rock commonly exist within a small area. If, for illustration, the state's sedimentary rocks are likened to almost-flat layers of cloth lying one atop the other, somewhat torn on the edges, then the Precambrian crystalline complex may be likened to a pile of tattered rags, wads and strings of various colors and fabrics, all jumbled up. Some are still intact and unrumpled enough to be recognized for what they originally were, but others are an enigma. By looking only at the outside of the pile or probing here and there, it is difficult to imagine what the original "garments" were, what pieces belonged together, how they were made, and in what order they were added to the pile.

The sedimentary formations do present problems also, as we shall see later, but the guidelines used to deduce their history are more reliable and more easily followed than those required to deduce the history of the ancient, greatly metamorphosed, long-eroded shield rocks. As explained in Chapter 1, fossils aid in telling relative ages of many sedimentary strata, and the strata as deposited are progressively younger from bottom to top. Even if the strata are tilted, warped or broken, or separated by erosional unconformities, the progression can still be followed; and even if they are dissected—cut into "fingers" or "islands" by erosion—they can usually be matched from one side of a valley to another. Some formations can be traced hundreds of miles. Drill holes into undisturbed sedimentary rocks penetrate a vertical sequence of strata that have a recognizable order.

But this is not the case in most of the cryptic, or puzzling, crystalline shield where rocks of widely different ages are juxtaposed, and where, only in some locations, do rocks present a layered or otherwise orderly

2. Wisconsin is famous for the variety and quality of its monument stone and dimension stone (cut, natural stone used for building), especially its granites, which combine strength, durability and beauty when polished. Several of Wisconsin's outstanding granites were used to decorate the interior of the state capitol in Madison, built during the early 1900s. They came from quarries in Marathon, Waupaca, Marinette, and Marquette counties, including those at Wausau, Montello, Waupaca, and Athelstane. (Attractive varieties of stone from other states and countries were also used in the capitol.)

sequence. Furthermore, fossils are rare to nonexistent in Precambrian rocks.

Here and elsewhere earth scientists have only recently begun to apply the concept of plate tectonics, which deals with the breaking apart of landmasses, the "drifting" of their divided parts to other locations, and the collision and reassembling of parts to form new continents. In the much-altered Precambrian bedrock of Wisconsin they are finding signs of such events in the past.

Because of the shield's complexity, only its main regional divisions and its most noteworthy events and features can be described in this summary. Data from early and recent investigations are brought together to give a composite picture, but admittedly much uncertainty and many blank spots exist.

The Precambrian Time Frame

Though the dating of Precambrian rocks is difficult, there are ways to estimate or measure the age of many of them. Intrusions and extrusions can tell the relative ages of adjacent rocks—that is, they tell which is younger and which is older. Intruding magma had to be younger than the rock through which it moved, and extrusions were younger than the surface upon which they flowed (fig. 2.3).

Some rocks can be dated by radiometric techniques, but these must be performed, and the results interpreted, cautiously, as errors can easily be made. In some cases the dates arrived at are only the time when the rocks were last molten or last subjected to intense heat or pressure, for recrystallization erased the earlier record of radioactive decay and gave the rock a new radiometric birthday of zero years. Much of the shield rock has presumably gone through heating several times since it originally formed, so a rock could well be older than the age given by radiometric measuring. Still, even the "reset" date of a rock can provide information about past geologic events and their timing. In years ahead radiometric dating will help greatly to clarify Precambrian history. Because only certain types of rock yield radiometric ages, and because the measuring process is costly and involved, only a limited number of Wisconsin rock specimens have been tested, and the dates are still too few to give a total picture. Ages of well over a billion years have been obtained from Precambrian rocks in different sites across central and northern Wisconsin, and it appears that some of Wisconsin's rocks are more than 3 billion years old. The northwest corner of the state is the youngest part of the shield in Wisconsin.

Since so much uncertainty exists regarding ages and relationships of the shield rocks, it has been difficult to make a chronological outline of Wisconsin's Precambrian history. In fact, as new data have been gathered over the years Precambrian terminology has been repeatedly changed. An explanation of some formerly popular terms may help in the reading of the older literature and maps.

At one time the Precambrian history was divided into the older or Archean and the younger or Algonkian periods. These names were then replaced by Archeozoic and Proterozoic, respectively, to conform with the nomenclature of the later periods ending in "zoic" (Paleozoic, Mesozoic and Cenozoic). Archeozoic included the older, more intensely metamorphosed Precambrian rocks, and Proterozoic the younger, less disturbed ones. But this division was not valid everywhere. In some cases similar rock formations in separated locations that had been classed as the same age were later found to be of quite different ages.

Some geologists chose to divide the Precambrian into units based on major mountain-building episodes known as *orogenies* (pronounced o-*rah*-jen-ees). One classification now commonly used divides the Precambrian, on that basis, into three parts. It speaks of the *Early, Middle* or *Late* Precambrian in referring to time, and of the *Lower, Middle* or *Upper* Precambrian in referring to the corresponding sequence of the stratified and intruded rocks. That classification is used in this book. (The old two-part classification does not translate directly into the three-part one. In some respects only, the Upper and Middle Precambrian are the counterpart of the Proterozoic, and the Lower Precambrian of the Archeozoic. The term "Archeozoic" fell into disuse, while "Proterozoic" has endured.)

Terminology is still far from standardized. Many geologists find certain classifications best suited to their purpose or their area. The relationships among some Precambrian divisions that have been commonly used are shown in figure 2.4. The U.S. Geological Survey and the Geological Survey of Canada now divide the Precambrian into two main sections, the Archean and Proterozoic. During the 1970s the U.S. Geological Survey used a W, X, Y, Z division where, in general, Precambrian W (older than 2,500 million years) corresponded to Lower Precambrian and the Archean; Precambrian X (2,500 to 1,600 m.y.) corresponded to Middle Precambrian; and Y (1,600 to 800 m.y.) and Z (800 m.y. to the base of the Cambrian) together corresponded to Upper Precambrian. That classification is no longer used. (See also the Geologic Time Scale, fig. 1.9.)

In most or all of Wisconsin the last few hundred million years or so of the Precambrian, as well as the start of the Cambrian, seem to be missing from the rock record (though some rocks in northwestern Wisconsin may have formed within that time). During that interval of more than 500 million years erosion predominated over the state, removing geologic evidence of what happened. The gap may seem unimportant, but

Rock Position and Sequence (used in this book)	Corresponding Time Group	Years Ago (estimate based on radiometric dating)	Former U.S. Geological Survey Classification	Current Classification of U.S.G.S. and Geological Survey of Canada
(Lower Cambrian)	(Early Cambrian)			
		— 600,000,000 —	Z	
Upper Precambrian	Late Precambrian	800,000,000 —		
			Y	
		— 1,600,000,000 —		Proterozoic
Middle Precambrian	Middle Precambrian		X	
		— 2,500,000,000 —		
Lower Precambrian	Early Precambrian		W	Archean
? ?	? ?	Earth formed	? ?	? ? ?

Figure 2.4 Comparison of some Precambrian divisions used in North America.

its magnitude can be appreciated when one considers that it covered a time span as long as all time from Early Cambrian to the Present. It is an understatement to say that much occurred in that interim.

The Early Precambrian

Of the opening of the Early Precambrian in Wisconsin or elsewhere little can be said, for what happened then is tied in with Earth's beginnings and formation of the initial crust (fig. 2.5). The first rocks undoubtedly formed from molten material. As those earliest igneous rocks were eroded, particles that would become shales, sandstones and conglomerates collected in low places. The sedimentary rocks they formed, and the present igneous rocks, were subjected to severe folding which possibly created mountains; underground intrusions of great masses of magma caused bulging and displacement of the rocks; and through these and other processes rocks became metamorphosed. Erosion kept wearing down the rough surface and ultimately exposed intrusive rocks that had been underground.

At the close of the Early Precambrian, Wisconsin was affected by an episode of upheaval and volcanism that probably formed mountains to the north of Lake Superior. This episode was called the Algoman Orogeny, named for the Algoma district of Ontario (fig. 2.5).

Wisconsin rocks that have been designated *Lower Precambrian (Archean)* are found in scattered locations on Wisconsin's shield (plate 1). Notable among these are rocks southeast of the Gogebic, or Penokee, Range[3] in Ashland and Iron counties. They occur too in other parts of northcentral Wisconsin; and in separated locations in central Wisconsin where the southern part of the shield is exposed.

(The term *Laurentian* was used in the past to designate granitic rocks intruded at the close of Early Precambrian, but it has been all but dropped. Geologists adopted it in the 1880s to refer to seemingly similar events that occurred considerable distances apart in Canada and northern United States, events that they assumed were simultaneous and related, but they did not then realize how complicated and inscrutable Precambrian geology was, and how difficult or impossible it was to correlate separated events. If used today, Laurentian refers to events at the start of the Algoman episode.)

The Middle Precambrian

A period of erosion led into the Middle Precambrian, a time that is characterized locally in Wisconsin by deposition of sediments and by volcanism, culminating in the Penokean Orogeny. Igneous intrusions penetrated existing rocks; lava poured out onto the surface; and at times volcanic explosions occurred, producing tuffs and breccias (volcanic rocks composed, respectively, of minute and large fragments).

Highlands were being reduced in elevation, and material eroded from them was deposited along with volcanic debris in low areas submerged by fresh water or sea water. These sediments included much sand and mud, which would harden into sandstone and shale and subsequently be changed into quartzite, slate, and other metamorphic variants. There were conglomerates too; those whose matrix was sand were metamorphosed to quartzite conglomerate, and if deformation was severe the matrix recrystallized and streaked gneiss resulted (fig. 1.7). Limestone was becoming conspicuous. By this

3. Described in chapter 3. Pronunciation: Go-gee-bick (with hard g's) and Pen-o-kee, both accented on the second syllable.

Rock Position and Sequence	Corresponding Time Group	Years Ago (estimate)	Certain Major Events
		600 million	? Precambrian/Cambrian boundary ?
			Mainly erosion, but continuing deposition in Lake Superior area
Upper Precambrian	Late Precambrian		Intrusions and faulting in Lake Superior area with continued sedimentation
		1.1 billion	Keweenawan Period: Accumulation of lava flows and sedimentary layers with contemporaneous sagging to form the Lake Superior syncline. Intrusions.
		1.5 billion	Forming of Wolf River Batholith
			Continued sedimentation, probably including that forming widespread quartzite
		1.8 billion	Penokean Orogeny: widespread folding, intrusions, and metamorphism
Middle Precambrian	Middle Precambrian		Deposition forming mainly sandstone and shale, minor amounts of limestone, and iron-formation, with considerable intrusion and extrusion of molten material
			Erosion and mild warping
		2.5 billion	Algoman Orogeny: metamorphism, folding, and intrusions
Lower Precambrian	Early Precambrian		Volcanic activity dominant
?	?	5 billion	? Earth was forming ?

Figure 2.5 A Precambrian chronology for the Wisconsin area.

time enough lime-secreting algae existed in the waters to leave calcium carbonate in quantity. The resulting limestone, when metamorphosed, became marble; presumably most marble that remains is deeply buried, but there are small outcrops of it like those along the Bad River southwest of Mellen and along the Marengo River in southeastern Bayfield County.

The Middle Precambrian sandstones were of various composition, ranging from fine to coarse grains and from almost exclusively quartz to an assemblage of many different mineral or rock particles.

Among the unrefined sandstones are the *graywackes* (the e is pronounced), which are dark-colored and poorly sorted. Along with light-colored quartz particles, graywackes contain much dark, fine-textured material that was formerly mud; pebbles of various sizes and composition; and angular bits of volcanic rock. This crude mixture looks as though it had been dumped rapidly into water without much sorting or abrasion. At that time there undoubtedly were many landslides and showers of volcanic ash, and much slope-washing and slumping under heavy rains as water flooded unrestrictedly over the bare land. Graywackes, now metamorphosed, are found in many parts of northern Wisconsin (fig. 2.6a).

At the opposite end of the sandstone gradation are lighter-colored rocks composed almost exclusively of well-sorted, well-rounded quartz grains (fig. 2.6b). Through washing, the fine muds and other non-sand materials were separated and removed. The "mature," sorted sand must have been worked in water for a long time, and may have gone through several cycles of deposition, breakup, and redeposition in order to have reached a state of such purity. Though now altered to quartzite, the rocks still show in places the slanting and crescent lines of cross-bedding by shifting currents, and ripple marks left by gentle waves on beaches and on other fresh sand deposits. Quartzites composed of strongly cemented, uniform beds of quartz sand are the most resistant rocks occurring in sizable outcrops in Wisconsin (fig. 1.7).

On the shield are many *monadnocks*—smoothed hills and ridges that have withstood erosion and are therefore higher than the surrounding, faster-eroding rock. Most of the monadnocks, and the largest of them, are quartzite. The ages of these various quartzites have not all been determined. It had been thought that most are of Middle Precambrian age, but now it is believed that some of those are Late Precambrian. More will be said of them in the next chapter.

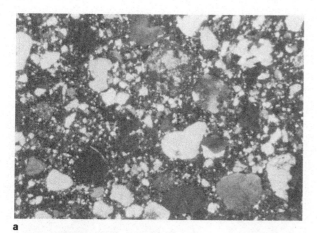

a

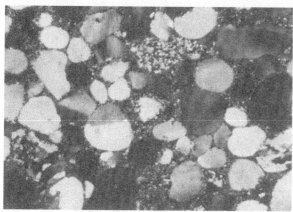

b

Figure 2.6 (a) A photomicrograph of a Middle Precambrian graywacke from near Hurley. The long dimension is 6 mm before magnification. It is a heterogeneous collection of particles which have not received smoothing by abrasion. **(b)** By contrast, the particles in this photomicrograph of a quartzite are smoothed and rounded by long abrasion, and through sorting most of the "dirty" fine particles have been removed. The long dimension is 6 mm before magnification. (Samples supplied by Robert H. Dott. Photography by Gordon Medaris.)

During the Middle Precambrian, in some parts of Wisconsin, chemical interaction of volcanic liquids and gases with sea water formed deposits that contain iron, zinc, copper, lead, silver, and gold.

A noteworthy event of this time was the deposition of *iron-formations* (iron-bearing strata) not only in Wisconsin but in shield areas around the world—in South America, Australia, India, China, USSR, Africa. There are various types of iron deposits, but the world's principal type is that of Middle Precambrian age characterized by a distinctive, layered composition which is known as "banded" or "Lake Superior type" iron-formation. The latter name came from the classic examples around that Lake—in the Gogebic and Menominee ranges of northern Wisconsin and northern Michigan, and in the Marquette Range of northern Michigan, and in the Mesabi and Cuyuna ranges of northern Minnesota. Still unanswered is the question

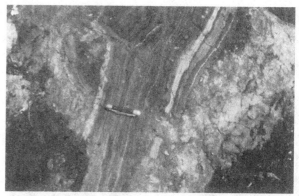

Figure 2.7 Iron-formation, with its parallel beds running perpendicular to the pocketknife. Basalt lies on either side. (W. O. Hotchkiss)

of what conditions prevailed during Middle Precambrian time to cause formations of this type to be deposited in many places around the world.[4]

A Lake Superior type iron-formation is a bedded sequence composed mainly of alternating layers of two kinds of rock, one rich in iron-bearing minerals, and the other largely chert (fig. 2.7). The layers, which may be a fraction of an inch to several inches thick, may be differentiated by their color, magnetism, or hardness. The deposits presumably were laid down in lagoons or in partly enclosed, shallow coastal waters where solutions bearing iron and silica (which would form chert) collected. The iron and silica were precipitated chemically and by living organisms, but in varying amounts. For some unknown reason, during certain times mainly silica was deposited—as a gel which would ultimately be converted to chert (fig. 2.8). This cherty layer would have a low iron content. During other times iron was deposited in significantly richer concentrations, along with much smaller amounts of silica, forming an iron-rich layer. So the layers built up with iron-rich and iron-poor beds alternating, accumlating in places to thicknesses of hundreds of feet.

Wisconsin's richest known iron-formation, that of the Gogebic Range, has already been mined in part. Mining has also been done in the Menominee Range in Florence County. Precambrian iron-formations that contain smaller concentrations of iron are known in Wisconsin, but the only ones that have been utilized thus far are those of the Baraboo Hills area, where

4. Certain of Wisconsin's rocks, including the iron-formation, used to be classed as "Huronian" in age but that term is no longer applied in Wisconsin. As first used, it designated rocks on the north side of Lake Huron in Canada. For a time the iron-formations of Wisconsin were thought to be the same age as those, but the most recent interpretation is that Wisconsin formations are younger. It seems no Huronian rocks are present in Wisconsin. At least, no rocks have been identified as such.

"Animikie" is a regional name that was given to some Middle Precambrian sequences of sedimentary rocks which include iron-formation in the Lake Superior area. It is an Indian word meaning "thunder" and was originally applied to rocks in the Thunder Bay area on Lake Superior's north shore.

Figure 2.8 Typical Indian projectile points chipped from chert of no definite source, from Manitowoc County. The earliest types are to the left, and progressively later ones, more delicately fashioned, are to the right. (By permission of the *Wisconsin Archaeologist* and John T. Penman, State Historical Society of Wisconsin)

Chert is usually light in color and so fine-grained as to appear smooth and uniform. Because it is extremely hard and breaks to a razor-sharp edge, chert and its darker counterpart, flint, were favorite rocks of Stone Age people, including the American Indians, who chipped them into spear and arrow points and tools. If chert is stained red, as some is in iron-formation, it is known as jasper. Chert occurs also as nodules in limestone and dolomite.

mining was done on a small scale in the past, and those near Black River Falls, where taconite deposits have been mined (fig. 2.9).

Taconite is a low-grade iron-formation, but it can be artificially enriched by crushing to release the iron-rich minerals, which are then magnetically segregated and formed into pellets having a higher iron content than the original ore (fig. 2.10). Taconite ore is so named because in the past it was erroneously thought to be the same age as similar rocks in the Taconic Range of New York.

In retrospect, it would appear that many Middle Precambrian sedimentary rocks, such as have been described, probably were broadly continuous formations when still horizontal. Then their continuity was broken by folding, by intrusions, and by large-scale tilting, lifting, and dropping of blocks of rock. Raised parts and weaker rocks were more vulnerable to erosion than rocks that were in depressions or that were more resistant. Now the once-continuous sedimentary rocks are preserved only in unconnected sites. Although certain rocks in separated parts of the shield may be similar to one another, it cannot confidently be said whether or not they previously belonged to the same formation.

The Penokean Orogeny was a widespread earth-deforming event across Wisconsin, Minnesota and Michigan during the Middle Precambrian (fig. 2.5). Folding and tilting of rock, faulting, and intrusions produced complex rock relationships. Highlands and

Figure 2.9 Districts in Wisconsin where iron ore has been mined.

Figure 2.10 Taconite pellets from the mine east of Black River Falls. Low-grade iron ore was crushed, worthless rock was removed, and the richer iron-bearing material was concentrated in this form for more convenient, economical shipment. (Ralph V. Boyer)

perhaps even some mountainous terrain were created across the northern Great Lakes region. This orogeny is dated at about 1.9 to 1.8 billion years ago. It is recorded by numerous intrusions. Radiometric dates have been obtained from some of them.

The close of the Middle Precambrian, then, left a zone of highlands which crossed northern Wisconsin. Erosion of these highlands would supply abundant rock-building sediments through Late Precambrian and into the Cambrian Period. The only part of Wisconsin that would still experience strong disturbance was the northwest corner. The rest of Wisconsin's shield area would feel only gentle warping and diminishing metamorphism, and then would become stable while erosion relentlessly wore down the surface.

The Late Precambrian

The Late Precambrian opened in Wisconsin with continuing erosion, deposition of sediments, and intrusions.

It may have been early in this period that some of the thick accumulations of clean sand that would later become quartzites were deposited over much of Wisconsin (fig. 2.5).

About 1.5 billion years ago a major intrusive event occurred in eastcentral Wisconsin, resulting in the *Wolf River Batholith* (figs. 2.11, 3.1). A *batholith* is the largest form of intrusion: an immense, broad body of magma from great depths slowly works its way upward into the crust, solidifying just below the surface, there to form a huge mass of igneous rock. A batholith is not a single body of rock, but consists of many smaller bodies that formed when the magma intruded at different times and places within a given area, which may include hundreds of square miles. Long afterward, if the rock above wears off, the top of the batholith may appear at the surface; this is happening here in Wisconsin. The large Wolf River Batholith (named for the Wolf River which flows across it) underlies an area of at least 3,600 square miles (9,300 sq km) whose outer limits are near Wausau, Stevens Point, Waupaca, Crivitz, and the southern boundary of Forest County (see also chapter 3).

That large intrusion was Wisconsin's last known important thermal event except for the *Keweenawan*

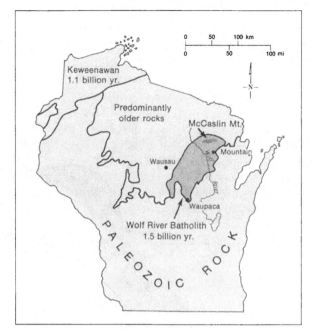

Figure 2.11 Location of the Wolf River Batholith in relation to other rocks.

episode in the Lake Superior area, which affected the northwestern corner of the state. The term Keweenawan (accent the third syllable) is applied both to the time of this activity, which was about 1.1 billion years ago, and to the area affected.

In general terms, the Keweenawan episode of the Lake Superior region is divided into the *Early Keweenawan,* a time of subsidence and deposition; the *Middle Keweenawan,* when widespread volcanism occurred; and the *Late Keweenawan,* when deposition continued (figs. 2.5 and 2.12).

Age	Rock Classification	Approximate thickness	Material		
Cambrian	Cambrian				
—— ? ——	—— ? ——	? ——— ?	? ——— ? ——— ?		
Cambrian or Late Keweenawan	Cambrian or Upper Keweenawan	4,300 feet (1,310 m)	Bayfield Group—sandstone	} Locally called "Lake Superior Sandstone"	
Late Keweenawan	Upper Keweenawan	17,500 feet (5,300 m)	Oronto Group—interbedded conglomerates, shales and sandstone.		
Middle Keweenawan	Middle Keweenawan	20,000 feet (6,000 m)	Lava flows interbedded with sedimentary strata, mainly sandstones and conglomerates		
Early Keweenawan	Lower Keweenawan	300 to 400 feet (90 to 120 m)	Sandstones and conglomerates		
/////// //////// /////// U N C O N F O R M I T Y ///////////////////// Period of erosion					
Middle Precambrian	Middle Precambrian				

Figure 2.12 Material accumulated in the Wisconsin section of the Lake Superior syncline. (These thicknesses are not all found at any one place.)

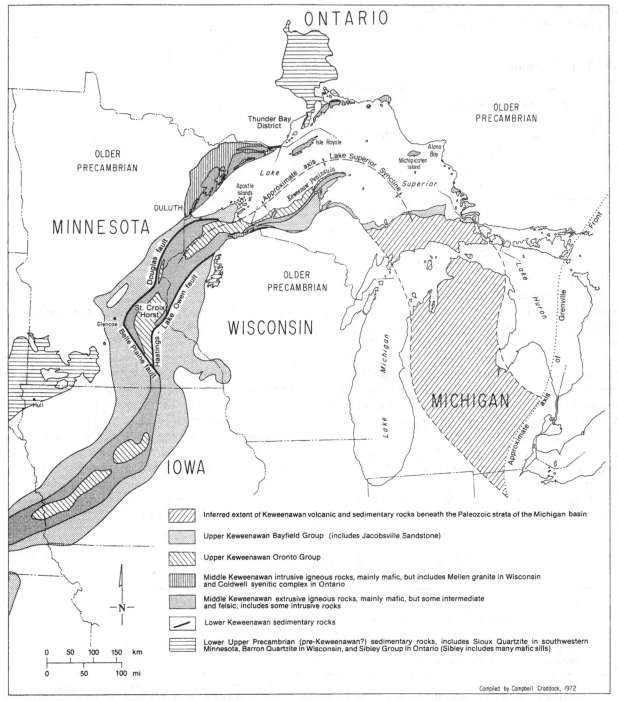

Figure 2.13 Generalized geologic map of the Lake Superior region showing the known and inferred distribution of Upper Precambrian rocks, compiled by Campbell Craddock. (Modified from C. Craddock, 1972, in *Geology of Minnesota: A centennial volume*, P. K. Sims and G. B. Morey, eds.)

Lake Superior, which came into being following Pleistocene glaciation, was of course not present in Keweenawan time. However, the nature of the material that accumulated there then, and the rock deformation that took place then, helped bring about the lake's existence and present shape. Prior to the existence of Lake Superior, water bodies of other configuration must have sometime occupied parts of the lowland.

At the start of Keweenawan time a long depression had begun to form in the Lake Superior basin area. It was part of a developing rift zone—a zone of faulting and volcanic activity—which was a major feature of the continent. This zone extended from Kansas through eastern Nebraska, Iowa, southeastern Minnesota, northwestern Wisconsin, and the Lake Superior basin into Michigan (fig. 2.13). The rift zone was a break in the shield, a split in the plate (section of the earth's

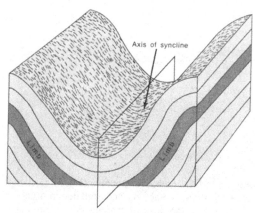

Figure 2.14 Diagrammatic drawing of a syncline, a downward bend of rocks.

crust) of which the shield is a part. The divided sections of the plate must have begun pulling away from each other, but full separation did not occur. Our focus will be on the part of this zone that includes Wisconsin—the western Lake Superior area—and it is in that part that the rocks of this period are best revealed. They are exposed in many places around the western end of the lake in Wisconsin, Minnesota and Michigan; but to the southwest, in Minnesota and beyond, they are covered by younger rocks, and in the other direction they dip under Lake Superior. Outcrops show that these rocks continue at least as far as the eastern end of Lake Superior, and subsurface detection indicates that they may curve and extend southward under the Paleozoic rocks of southern Michigan.

Along this break in the crust magma rose from below, and the surface began to slowly sink. The newly forming depression assumed the shape of a *syncline,* which is a bedrock trough—a long, narrow section of the crust that has been lowered. The lowering is generally down-folding, but may, as in the Lake Superior area, include considerable faulting also. Here the trough included what is now the lake basin and its shore areas (fig. 2.14).

Northwestern Wisconsin lay along the southeast limb, or side, of this syncline, so as sagging took place the affected rocks there (as well as in the Keweenaw Peninsula to the northeast) tilted down toward the northwest, into the trough. On the opposite, Minnesota side of the syncline (including Michigan's Isle Royale), rocks were accordingly tilted downward the other way, to the southeast.

The synclinal trough in the Lake Superior area kept sagging slowly during the Keweenawan period, and as it did so sediments kept washing into it from bordering higher land. They accumulated there in horizontal layers, which became sedimentary rock. These horizontal rock layers in time would also become tilted inward as the trough sank further. For millions of years the sinking continued. Ultimately the lowest part of the

syncline would reach a depth of about 50,000 feet (15,000 m)—almost ten miles. (Fig. 2.15.)

The Lake Superior syncline is asymmetrical. Its southeast limb on the Wisconsin side is tilted steeply, while the opposite northwest limb along the Minnesota side is tilted much less. The trough's axis, or line of greatest depth, lies toward its southeast side, running northeast-southwest between Isle Royale and the Keweenaw Peninsula of northern Michigan (which gave the name to the area and time), across northern Ashland County, through southern Douglas County, and into northern Burnett County. (Figs. 2.13 and 2.16.)

In the *Early Keweenawan,* when the depression was only beginning to form, sands and gravels were deposited there in a shallow-water environment to a thickness of 300 to 400 feet (90 to 120 m), sinking as they accumulated. Those sands have since been changed to quartzite.

Then during *Middle Keweenawan,* volcanism increased, and one of the world's great outpourings of lava occurred. Though the following discussion centers on the western Lake Superior area, volcanic and diastrophic (folding and faulting) events similar to those of this area were likely occurring throughout the rift zone.

Fissures opened along fractures, and from them lava poured out over the lowland time and time again. For the most part it was basalt, a dark heavy lava, but some of the flows were rhyolite. These lava emissions are considered extraordinary because they were not just small extrusions, but voluminous floods, and there were hundreds of flows. Though the lava floods in total were copious, they took place gently, not explosively.

The lavas came intermittently from various fissures and spread in all directions over the lowland, coalescing, burying large areas, and sometimes flowing into water bodies that were there. Each new emission, covering previous ones, created a new gently sloping or horizontal surface. Sometimes the flows came in such close succession that little weathering, erosion, or deposition could take place in the interval between them; but enough time often elapsed between flows for layers of sand and gravel to be deposited, so sedimentary strata of sandstones and conglomerates were interlayered among the lava flows. The sediments seem to have been deposited by streams flowing into the basin and meandering over the surface, or building aprons or deltalike fans of alluvium (stream-deposited material); some sediments accumulated in lakes.

The basaltic lava came up fiery hot, and cooled and thickened as it spread. It cooled fastest at its top and bottom surfaces where it was in contact with the air or with the cool land beneath, or occasionally with water. These surfaces, especially the upper one, often solidified before all bubbles of gas had escaped. The trapped bubbles caused cavities, or vesicles, which averaged

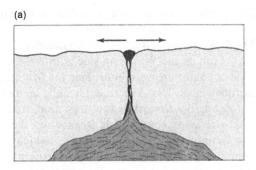

(a)

Rifting in earth's crust allowed magma to rise to surface as lava.

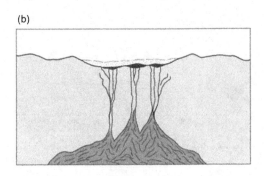

(b)

Outpouring of lava increased. Sediments were carried into area by streams. Sagging occurred due to greater weight at surface, loss of magma below, crustal disturbance, or a combination of these causes.

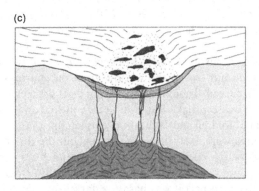

(c)

Syncline continued to sag as more layers of lava and in-washed sediments accumulated on the surface, and as the rift widened.

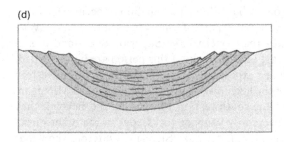

(d)

Hundreds of layers of lava and sediments built up and synclinal sagging continued. Upper layers are more horizontal than lower ones which have experienced sagging longer.

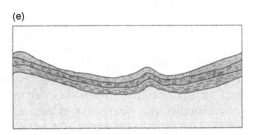

(e)

Folding occurred in parts of the syncline.

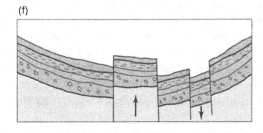

(f)

Faulting also occurred in many places within it, with sections rising or dropping. During this time, of course, fault edges were not sharp as shown, for erosion was wearing down higher parts, and lower parts were filling with sediment.

Figure 2.15 Diagrammatic illustration of Keweenawan events in the Lake Superior region.

about a quarter of an inch in diameter (fig. 2.17). Some vesicles were spherical; others were elongate and irregular in shape. They usually mark the top of a flow and so aid in distinguishing individual flows.

These cavities are significant for another reason. They slowly became filled with materials that were carried into them by hot water and gases associated with the volcanism—generally calcite or quartz, but locally copper or silver or both. The holes were gradually coated layer by layer, becoming smaller and smaller hollows. Not all were filled completely. Those that were filled are called amygdules (fig. 2.18).

A quartz filling consisting of a number of layers of different colors is an agate. Millions of years later when the lava was exposed and disintegrated, the agates, which were more resistant than the matrix that held them, were released. These semiprecious stones continue to be washed up on Lake Superior shores. Many

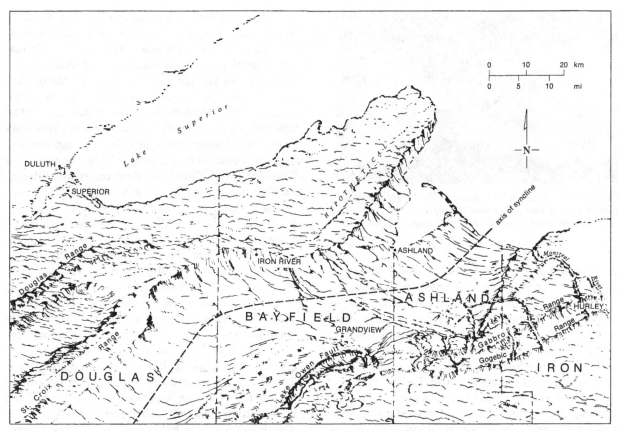

Figure 2.16 A physiographic diagram showing Precambrian structures that influenced topography in the Keweenawan and Gogebic Range area as envisioned by H. R. Aldrich in 1929. Of course, the bedrock was heavily glaciated and is covered by drift and other material, so the land surface does not look like this. Names of some features have been added. (From H. R. Aldrich, *The Geology of the Gogebic Iron Range of Wisconsin*, Bulletin 71, Wisconsin Geological and Natural History Survey, 1929)

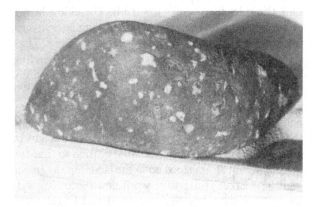

Figure 2.18 Basalt containing amygdules, from Douglas County south of Superior. (Ralph V. Boyer)

Figure 2.17 The little spherical holes, or vesicles, in this basalt were caused by bubbles of gas trapped in the lava when it cooled. Leaves provide a scale of size. (George Knudsen, Wisconsin Department of Natural Resources)

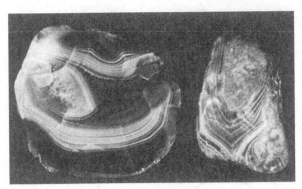

Figure 2.19 Lake Superior agates. The larger one is 2 inches wide. (Geology Museum, University of Wisconsin-Madison. Lawrence D. Lynch, photographer)

are broken, but are worn smooth by being worked in water. They are sought by rock collectors and are used in jewelry because their colorful, semitranslucent, concentric layers of brown, red, white, and tan have interesting designs, and because they are hard enough to take and hold a polish. Many agates found on beaches are a quarter to a half inch wide. Large agates may be up to three inches (8 cm) wide (fig. 2.19).

Copper was deposited not only in vesicles, but also in openings in the sedimentary rocks. Many of Wisconsin's Keweenawan rocks contain small amounts of copper.

As heavy magma from beneath the surface was transferred above ground as lava, and upward as intrusions, there was a resulting weight increase above and perhaps less support below. As the trough's center sank the sides tilted down toward it. The oldest layers eventually became more steeply tilted than later layers, for they went through the tilting process longer. They were at the bottom of the "stack" and were bending into a canoe shape while new layers were still forming horizontally at the top of the shallow basin. In other words, the younger and higher the layers, the less they had sagged.

The upturned, exposed edges of the Keweenawan strata were subjected to differential erosion so that the more resistant strata stood up in the landscape as ridges and the weaker strata were worn down into long valleys. (Glaciation has since modified and largely covered the surface, so the ridges are evident only in certain places.) The sequence of alternating lava beds and sedimentary strata on the Wisconsin-Michigan side is essentially a mirror image of the Minnesota side. The strata, before erosion, probably covered a wider area than they do now, extending beyond the Lake Superior basin. They have been preserved in the basin area because that lower position made them less subject to erosion.

During the Middle Keweenawan the lava flows and interlayered sediments in the Wisconsin section accumulated to a thickness of more than 20,000 feet (6,000 m). A fourth to a third of that layered thickness is sedimentary rock and the rest is lava. As the Middle Keweenawan volcanic episode got under way most of the material being laid down in the basin was lava, but as time went on there were progressively fewer lava flows in relation to sedimentary deposits. The sedimentary layers increased in thickness as intervals between lava flows lengthened. Finally the flows ceased and Wisconsin's last great period of volcanism ended.

In the *Late Keweenawan* deposition of sediments continued over the Lake Superior lowland, building up numerous layers. A 5,000-foot (1,500-meter) covering consisting mainly of coarse conglomerates (Outer Conglomerate) was followed by 500 feet (150 m) of shaly material (Nonesuch Shale) and 12,000 feet (3,600 m) of fine-grained sand with some mud (Freda Sandstone). The resulting sandstones and shales are mostly reddish. Ripple marks are common, showing that deposition often occurred in shallow water. This great thickness of strata constitutes the *Oronto Group* (figs. 2.12, 2.13). Some intrusions of magma continued into Oronto time. During deposition the sagging of the syncline continued, with subsidence and sediment accumulation nearly in balance, as before. These upper strata were not down-warped as much as the Lower and Middle Keweenawan ones.

Due to lateral compression from the sides, smaller folds formed within the larger fold of the syncline—in the Oronto beds and in older Keweenawan rocks in some localities. The axes of these smaller folds are approximately parallel to the syncline's main axis.

Overlying the Oronto Group are many nearly horizontal layers of sandstone composed of materials ranging from coarse grits to fine, shale-size particles, and they too are generally red. In lower-to-upper sequence they are the Orienta, Devils Island, and Chequamegon formations. Collectively they comprise the *Bayfield Group*, and have a thickness of about 4,300 feet (1,310 m) (figs. 2.12, 2.13). They are broadly, but only slightly, synclinal, and are not deformed as were the underlying rocks. Their best-known area of occurrence in Wisconsin is in the Apostle Islands and in Bayfield Peninsula (for which the group is named), and westward along the lake shore to the city of Superior— that is, north of the Douglas Fault (described in the next section). The Bayfield rocks extend into Minnesota, where they are known as the Hinckley Sandstone and the underlying Fond du Lac Formation. They and the Oronto Group may extend as far south as Kansas in the subsurface.

Summarizing the thickness of Keweenawan strata in the Wisconsin section of the Lake Superior depression, it is seen that Lower Keweenawan sediments, known from only a few sites, are on the order of a few hundred feet thick; Middle Keweenawan strata, which consist of lava flows interbedded with sedimentary strata (mainly sandstones and conglomerates) amount to about 20,000 feet; the overlying sedimentary Oronto Group amounts to 17,500 feet; and the Bayfield Group to 4,300 feet. But all this thickness is not found in any one place.

In Wisconsin the Oronto and Bayfield groups together are locally referred to as "Lake Superior Sandstone." The contact zone between them is nowhere clear. It is not certain whether their age is Late Precambrian or Cambrian. No fossils or datable minerals have been found in them to resolve the question. At this time the Oronto and Bayfield groups are generally considered Upper Precambrian (Upper Keweenawan), though some strata may be Cambrian. This region's sedimentary rock record ends there. The major faults had already occurred in the Lake Superior area by the time the upper Bayfield sandstones were deposited, but small displacements continued through Paleozoic time.

Some Features of the Rift Zone

The rift zone is concealed along much of its length by a cover of younger rock and glacial drift; it has been located by drilling and by sensitive instruments that detect variations in gravity strength and magnetism from place to place, and thus identify rocks of unlike density and composition. Along this rift zone is a belt of higher-than-average gravity values, known as the *Midcontinent Gravity High*. It is the most prominent gravity anomaly (deviation from normal values) known in the United States.

As we have seen, a major break and separation in the crust, which began in Early Keweenawan time, permitted magma to move upward and out upon the surface, and allowed the surface to sag. Severe faulting broke up the syncline, and blocks of the earth's crust were raised or lowered. In northwestern Wisconsin and eastcentral Minnesota the largest of many faulted features that disrupt the orderly layered arrangement of the syncline is a feature called the *St. Croix Horst*, which appears to be an uplifted crustal block (fig. 2.13). A *horst* is a large section of rock, bounded by parallel faults, that stands higher than its surroundings (fig. 2.2).

The St. Croix Horst was named for the St. Croix River which, along with its tributaries, flows over its general area. Being within the Lake Superior syncline, it is composed mainly of layered lava flows and interbedded sedimentary strata which have the synclinal warp. Because it contains part of the axis, or lowest line (before uplift), of the Lake Superior syncline, it is termed an "axial horst." It is tens of miles wide and shows a relative uplift of about 10,000 feet (3,000 m). But while it was slowly uplifted over millions of years it probably appeared no more than a minor rise on the landscape. Being constantly eroded, it contributed sediments to the filling of the lowlands along its sides. It may have been totally covered by Late Keweenawan (or Cambrian) sandstone in Wisconsin, but if so that sandstone was stripped off its higher parts, and now remains only in the low axis of the syncline.

The St. Croix Horst is bounded lengthwise by the *Douglas Fault* and the *Lake Owen Fault*, between which it was apparently raised (fig. 2.13). It should be mentioned that a minority of geologists believes these breaks are not faults, but merely unconformities, in which case the feature would not be a horst. The breaks—whether faults or unconformities—are observable in only a limited number of exposures and are indistinct. The picture of what actually happened is further blurred by glacial drift. In any case, the Douglas and Lake Owen faults are not to be considered just single faults, but are zones of numerous faults that lie parallel to the main one. Also, many short faults cross these fault zones at approximately right angles.

The Douglas Fault enters Wisconsin from Minnesota and runs across northern Douglas County into Bayfield County. Beyond there, if it continues, it is lost under Late Keweenawan sedimentary rock, glacial drift, and Lake Superior. Along this fault the raised northwest side of the St. Croix Horst is a ridge known as the *Douglas Copper Range* or just the *Douglas Range*.

The Lake Owen Fault, along the southeastern side of the horst, is a continuation of the *Hastings Fault* of Minnesota. It enters Wisconsin at St. Croix County's southern border at the Mississippi River and runs northeast along an irregular line to north of Spooner, to Lake Owen in southcentral Bayfield County, and on to the eastern Bayfield County border. It is not known north of the Gogebic area, but it may connect with the *Keweenaw Fault* which parallels the Gogebic Range on the northwest, following the trend of the syncline. From the boundary of Bayfield and Ashland counties the Keweenaw Fault runs northeast, passing just north of Mellen, and continuing into Upper Michigan along the length of the Keweenaw Peninsula.

The St. Croix Horst is terminated on the southwest by the *Belle Plaine Fault* in Minnesota (fig. 2.13). A similar horst feature lies south of there. The northeastern extent of the horst is unknown. It may end in northwestern Wisconsin or, if the Keweenaw Fault marks its edge, it may continue into eastern Lake Superior.

Some geologists believe that the lowland south and east of the cities of Superior and Duluth is a *graben*.[5] A graben is the opposite of a horst—a section of rock that *dropped* between parallel faults (fig. 2.2). They see this lowland as being bounded on the southeast side by the Douglas Fault (Douglas Range), and on the northwest side by the high escarpment (which they consider a fault line) that runs along the northwestern shore of Lake Superior at Duluth. Other geologists do not view that escarpment as a fault line, because they see no geophysical evidence for a fault, and so they do not consider the lowland to be a graben.

The largest local intrusions in the Keweenawan area of Wisconsin were along the northwestern side of the Keweenaw fault, which provided an outlet for up-welling magma. A large intrusive body of granite lies northwest of Mellen, and still-larger gabbro intrusions lie west and east of it. These cooled and hardened far underground in Middle Keweenawan time and are now exposed. Some of the gabbro that was quarried near Mellen was marketed as "black granite" (fig. 1.5).

From the vicinity of Mellen northeast to Michigan these intrusions, along with the up-turned edges of northwest-dipping lava beds, have resulted in a highland belt sometimes called the Keweenawan Highlands. Lava beds southeast of the intrusions form the highest part of the belt—the *Gabbro Range,* which is about 3 or 4 miles (5 or 6 km) wide and parallels the Gogebic Range (Penokee Range) along its north side. The Gabbro Range is one of a series of ridges which decrease in elevation to the north, toward the Lake Superior sandstone lowland. An escarpment marks the break between the lavas and the lower sandstone (fig. 2.16, and plates 3 and 4).

Looking back through the Precambrian Era we see that it is an exceedingly long, largely obscure portion of Earth's time. But lately, because of increased numbers of investigators and revelations brought about by modern research techniques, some Precambrian events, heretofore unperceived, are looming up out of the once-vacant past.

5. "Graben" is a German word for "ditch" or "trench."

3

Precambrian Wisconsin, by Regions

We have traced Wisconsin's Precambrian geology through the course of time, describing the general sequence of events, as currently known, that made this part of the shield what it is. Now we shall look at individual areas within the shield—defining the area, giving information about its geologic makeup, and, where pertinent, noting the influence of geology upon local geography and history.

The regional divisions of the Precambrian shield used here are shown in figure 3.1. They are based in large part upon those of the 1970 report by Dutton and Bradley[1] that contains the most recent, comprehensive, statewide compilation of Wisconsin's Precambrian geology, mapped in detail. That report is a summary of information found in publications and files available at that time and in unpublished data of the Wisconsin Geological and Natural History Survey, including geologists' field notes, well-drilling records, and aeromagnetic and gravity surveys. Since that report was prepared, more data have been acquired and more of the Precambrian area has been mapped, so in the figure 3.1 map some of the Dutton-Bradley divisions, or areas, are modified. The boundaries of the areas are subject to future revision, and have been generalized here to fit a scale smaller than the original. Different researchers using different bases of classification make other regional divisions. (Descriptive information about the areas comes not only from Dutton and Bradley, of course, but from many sources and observations.)

The Keweenawan Area

The Keweenawan Area, youngest of the Precambrian regional divisions in Wisconsin, occupies the northwest corner of the state (fig. 3.1). As we have seen in chapter 2, it was involved in the synclinal-volcanic-rifting activity of the Lake Superior basin region in Late Precambrian time. Its southeastern boundary extends from just below the "nose" of the Indian Head profile, northeast along the edge of younger Cambrian rock,

irregularly along the Lake Owen Fault (plate 1 and fig. 3.2), and finally along the Gogebic Range Area. The center of the "nose" is overlain by Cambrian rock and therefore is not included in this Precambrian division. Where the Keweenawan Area meets Lake Superior it continues under the lake. On the west it extends into Minnesota.

Although Keweenawan bedrock is largely masked by glacial drift, the southwest-northeast structural orientation—produced by the alignment of the syncline and major faults—is manifest in the topography when viewed in broad perspective. Western Lake Superior is oriented that direction, as are Bayfield Peninsula and the bays on either side of it. The Douglas Range, a conspicuous topographic feature, also exhibits this trend through most of Douglas County. The Gabbro Range too runs southwest-northeast, paralleling the Gogebic Range (fig. 2.16).

Had glaciation not covered the Precambrian rock, more of this grain of the land would be apparent. On the upbent flanks of the syncline (where they are not covered by Oronto and Bayfield sedimentary rocks) lava strata alternating with weaker, sedimentary strata would probably have produced a topography of alternating parallel ridges and valleys trending generally southwest-northeast. The edges of uptilted lava beds would be ridges, and the weaker, more easily eroded sedimentary layers between them would be valleys. But while glacial deposition largely concealed these linear lava ridges, glacial erosion accentuated the southwest-northeast grain in other ways.

When the first of several Pleistocene ice sheets spread down from Canada the Lake Superior basin was presumably not a lake basin, but more likely the drainage basin of an east-flowing river. The low main river valley and its tributaries would have been natural

1. Carl E. Dutton and Reta E. Bradley, *Lithologic, Geophysical, and Mineral Commodity Maps of Precambrian Rocks in Wisconsin* (Washington, D.C.: U.S. Geological Survey, 1970).

Figure 3.1 Main regions of exposed Precambrian rocks in Wisconsin, divided into areas. (Modified from Dutton and Bradley)

avenues for the expanding ice sheet to follow. Thick, heavy ice—like that flowing from Greenland and Antarctica now—bulged southwestward through that low area. It dug out the relatively weak sedimentary rocks at the top of the syncline, as well as lava that was with them. The ice was divided by the highland that is now Bayfield Peninsula, and it scraped out weaker rocks along the sides of the peninsula. Thus this ice sheet and later ones deepened the Lake Superior basin and its two southwestern bays, leaving Bayfield Peninsula standing between the bays.

Most of the *Apostle Islands* are the ice-scoured tops of sandstone hills, now partly submerged, where the tip of the eroding peninsula bore the brunt of the ice pushing down from the northeast (fig. 3.3). Probably before the ice came most of the islands were part of the peninsula, which was longer then.

The islands vary considerably in elevation. Some of the smaller ones rise no more than 20 feet (6 m) above the lake, while the highest, Oak Island just off the peninsula's tip, rises more than 500 feet (150 m) above it. The southernmost, Long Island, is a low ex-

44

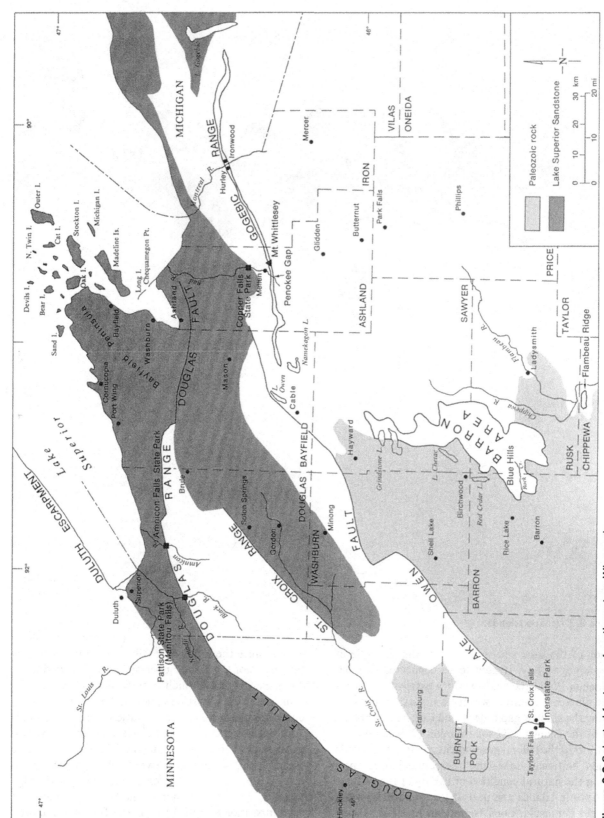

Figure 3.2 Selected features of northwestern Wisconsin.

45

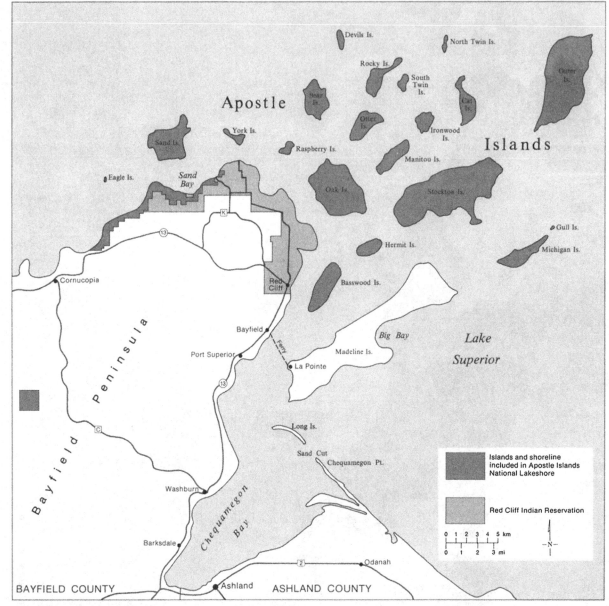

Figure 3.3 Apostle Islands.

tension of the sand spit that protects the east side of Chequamegon Bay. It used to be connected to Chequamegon Point, the spit's western extremity. Storms in the nineteenth century washed away part of the spit, causing the shallow gap called Sand Cut. The total land area of the islands is 77 square miles (200 sq km).

In 1970 the federal government created the Apostle Islands National Lakeshore recreational area, preserving the natural condition of twenty of the twenty-two Apostle Islands and part of the adjacent shore of Bayfield Peninsula. Long Island and Madeline Island were the two islands not included. Madeline is the largest of the islands (14,904 acres), and much of it was already privately developed when the recreational area was acquired. There are plans now to include Long Island also.

Because the ice sheet easily dug up the sedimentary rocks of the Lake Superior basin it carried a copious load of drift, which it deposited to the south as it melted. Heavy drift was piled on Bayfield Peninsula, adding substantially to its elevation. The rough highland running along the spine of the peninsula, known as Bayfield Ridge, rises to heights of about 600 to 800 feet above Lake Superior (whose elevation above sea level is about 600 feet). Topographically Bayfield Ridge appears to be an extension of the Douglas Range, but its composition is quite different. Drift several hundred feet thick accounts for most of its height (plates 2 and 3). More is said of the effects of glaciation in chapter 6.

In spite of the leveling and covering-over done by glaciation in the Keweenawan Area, some lava ridges trending southwest-northeast are visible at escarpments, road cuts, gorges, and waterfalls. The ridges are generally parallel to one another and dip toward the axis of the syncline, usually at an angle of less than 30 degrees. Where undisturbed by faulting, their eroding edge faces the outside of the syncline and has a relatively steep face—often an escarpment—whereas the top of the bed which slants down into the syncline's trough has a gentler slope. Good exposures of lava beds are seen at St. Croix Falls and along the St. Croix and Douglas "copper ranges" (so described because some copper is found in the lavas).

The *St. Croix Range* is the edge of lava formations in southern Douglas County, bordering the sandstones (Oronto Group) which lie to the southeast. The range runs southwest-northeast along the northwest side of the upper St. Croix River, continuing on to just northeast of Solon Springs, roughly parallel to the Douglas Range (fig. 3.2).

The better-known *Douglas Range* is the edge of the uplifted St. Croix Horst along the Douglas Fault (figs. 2.2, 2.13). It consists of a series of ridges along a line that runs near the towns of Patzau, South Range, Wentworth, Poplar, and Maple, and continues eastward. The lava formations of the range dip southeast into the syncline trough at angles of 35 to 40 degrees. Its northwest-facing escarpment overlooks the Lake Superior Lowland, between it and Lake Superior, where lava formations at lower elevations are covered by sandstones (Bayfield Group) and glacial-lake sediments (plate 4). (The Douglas Range is sometimes referred to locally as the "South Range" because it is south of Lake Superior. The escarpment running northeast from Duluth along the northwest shore of Lake Superior used to be viewed as its northern counterpart and was called the "North Range.") The Douglas Range has a local relief of about 350 feet (100 m). Its escarpment is not abrupt; it drops about 160 to 300 feet in a mile. Pattison and Amnicon Falls state parks are on this escarpment. Their settings are not only scenically attractive, but geologically revealing (fig. 3.2).

Pattison State Park, about 10 miles south of the city of Superior on State Highway 35, contains Wisconsin's highest waterfall, Manitou Falls (Big Manitou Falls) (fig. 3.4). There the Black River, a north-flowing tributary of the Nemadji, cascades in a fall of 165 feet (50 m) over the escarpment of resistant lavas of the Douglas Range to the plain below. (This river is not to be confused with the larger Black River in westcentral Wisconsin.) The Douglas Fault between the sandstone of the Lake Superior Lowland and the raised block of lava formations is difficult to locate, but on Highway 35 it is approximately at the north boundary of the park. At the falls the walls of the gorge are lava. Below, the river flows for about a mile in a canyon cut 100 to 170

Figure 3.4 Big Manitou Falls, Wisconsin's tallest waterfall, in Pattison State Park south of Superior. The north-flowing Black River tumbles 165 feet (50 m) over an escarpment of resistant basalt. (Wisconsin Department of Natural Resources. Photographer, Dean Tvedt)

feet deep in younger sandstone. The river took this route when the last ice sheet withdrew, and it has eroded the gorge since then. A smaller waterfall in this park is Little Manitou Falls, upstream and about a mile south of Big Manitou Falls. There the river drops 31 feet (10 m) over a lava ledge (fig. 3.5). Also seen in the park are Interfalls Lake, created by a dam above Big Manitou Falls, and several former glacial-lake beaches at the foot of the escarpment which were made by the lake when it stood at different levels during the recession of the ice sheet. A glacial lake that preceded Lake Superior, and that was higher than the lake is now, reached inland up to the escarpment and then cut successively lower shorelines as it drained down through various stages to its present level (fig. 6.33). The park of more than 1,300 acres is named for Martin Pattison of Superior, original owner of part of the park.

Amnicon Falls State Park, more than 800 acres in size, is a few miles southeast of Superior, near the junction of U.S. highways 2 and 53. It features many small falls where streams tumble over lava beds. On the east bank of the Amnicon River, under the park's main falls, can be seen a section of the Douglas Fault along which

Figure 3.5 Little Manitou Falls upstream from Big Manitou Falls drops over the steep face of an outcrop of lava. (Wisconsin Department of Natural Resources. Photographer, Dean Tvedt)

lava flows a billion years old appear to have been displaced northward over younger red sandstones of the Lake Superior Basin (fig. 3.6).

Many other streams create falls and rapids as they flow over this escarpment to the Lake Superior Lowland.

The *Lake Superior Lowland* along the lake shore of the Keweenawan Area is a relatively low region of Late Keweenawan and perhaps-Cambrian sandstones, with lesser thicknesses of conglomerates and shales. (The Nonesuch Shale may contain oil and gas.) Those sedimentary rocks have been covered by reddish glacial-lake sediments—fine sands and clays. They were deposited on the floor of the enlarged lake that preceded Lake Superior near the end of the Ice Age. The boundary of the Lake Superior Lowland is the highest abandoned beachline of that glacial lake which washed against Bayfield Ridge, the Douglas Range, and other Precambrian highlands (plates 2 and 4).

The Lake Superior Lowland consists of two broad areas, west and east of Bayfield Ridge, focusing on the cities of Superior and Ashland respectively; and of a connecting narrow lakeshore strip along Bayfield Peninsula. The western area fronts on the lake shore from the Minnesota border to near Cornucopia, and the

Figure 3.6 Amnicon Falls, one of the many waterfalls along the Douglas Range where streams cascade over the lava escarpment to the Lake Superior Lowland. Here the Amnicon River crosses the Douglas Fault. Lava rock above moved north over the lower, weaker sandstones, which are in the foreground where the men are standing. (Richard W. Ojakangas)

eastern area from west of Ashland to the Montreal River at the Michigan boundary. The Lowland is most level at Superior and Ashland. Elsewhere, though generally flat, it has been dissected by streams and gullied, so that in places it is hilly. It slopes toward the lake from interior elevations of about 1,100 feet (335 m) above sea level to shoreline elevations of about 620 or 650 feet where there are wave-cut banks and bluffs 20 to 50 feet high, or to marshes and beaches at lake level. Geographically the Apostle Islands are considered part of the Lake Superior Lowland. (Fig. 3.7)

Red and brown sandstones from the Keweenawan Area, including the Apostle Islands, were used in the construction of hundreds of residential, commercial and public buildings in many states between 1860 and the turn of the century. Buildings made of the brown sandstone are commonly referred to as "brownstones." In Wisconsin some of the buildings constructed of Lake Superior Sandstone have been torn down because of age, but quite a few remain, and some are being preserved for their historic value. (Fig. 3.8)

Figure 3.7 Wave-eroded sea caves in horizontally bedded Devils Island Sandstone (belonging to the Bayfield Group). Devils Island, Apostle Islands. (Apostle Islands National Lakeshore)

a

b

c

d

e

Figure 3.8 Red and brown Lake Superior Sandstone was a choice, widely used building material prior to the steel-and-cement era. Some of the many buildings constructed of this rock in Wisconsin are shown here. **(a)** St. Paul's Episcopal Church at E. Knapp and N. Marshall streets in Milwaukee, built of Lake Superior Sandstone in 1883. (Richard W. E. Perrin) **(b)** Forest Home Cemetery Chapel, Milwaukee, built in 1890. (Richard W. E. Perrin) **(c)** Washburn State Bank, Washburn. (Ralph V. Boyer) **(d)** St. Louis Catholic Church, Washburn, built in 1902. (Ralph V. Boyer) **(e)** City Hall in Ashland, formerly the post office, built in 1892. (Ralph V. Boyer)

Figure 3.9 The St. Croix River gorge in Interstate Park reveals tilted lava beds on the left side and large jointed blocks on the right. Much of the cutting of the gorge was done as the last ice sheet melted and huge volumes of water poured from it, some of it following this spillway to the Mississippi River. In early days this river was part of the much-traveled Bois Brule-St. Croix water-route trail, the shortest connection between the Great Lakes and the Mississippi. (Wisconsin Department of Natural Resources. Photographer, Robert Espeseth)

Just below the city of St. Croix Falls in western Polk County the St. Croix River has cut a gorge into Keweenawan lava flows (fig. 3.9). French explorers who traveled by canoe named this gorge the "Dalles of St. Croix." Because of the joints, or cracks, in the lava, rocks fall off in large blocks and the river takes an angular course. Steep walls rise about 100 feet (30 m) above the water, with still-higher prominences, and show a cross-section of tilted lava beds. The gorge was cut as the last ice sheet retreated and sent meltwater south through this outlet. The St. Croix River then was much larger than it is now, and had greater erosive power. Bowl-like potholes were rounded into the rock by sediment and stones that swirled in eddies in the bed of the river as it carved its gorge (fig. 3.10). The largest potholes are on the Minnesota side of the river. The geologic features and scenic beauty of the gorge area led to the establishment of Wisconsin's first state park at that site in 1900. This is *Interstate Park,* administered partly by Wisconsin and partly by Minnesota, for the river forms the state boundary there. The park, more than 1,300 acres in size, is now included in the Ice Age National Scientific Reserve.

Remaining parts of several terraces indicate where the St. Croix River formerly flowed at higher elevations: one terrace is at the level of the main street of St. Croix Falls; a lower terrace is upstream from the town along the river road; the main part of the town of Taylors Falls, Minnesota, is on a still-lower terrace. The

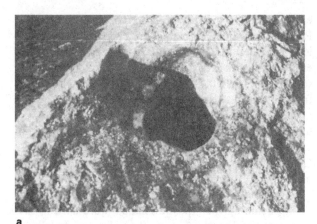

a

b

Figure 3.10 (a) This pothole at Interstate Park is about a foot across and over a foot deep. There is usually water in it. Potholes like this were ground out by stones and grit swirled around by plunging or eddying water. (George Knudsen, Wisconsin Department of Natural Resources) **(b)** Rounded stones that did the grinding in the larger potholes in the park are seen here. The largest one (basalt) is bushelbasket size. The small light-colored one (granite) is bowling-ball size. (George Knudsen, Wisconsin Department of Natural Resources)

river was a main thoroughfare in the days of river travel. The towns of St. Croix Falls and Taylors Falls are located where canoeists portaged around steep rapids. Damming of the river caused the falls to be submerged.

In the complex eastern part of the Keweenawan Area are some notable faults. A major one is the Keweenaw Fault which passes north of Mellen and on up Michigan's Keweenaw Peninsula. The shorter Bad River Fault crosses the Gogebic Range in a nearly south-north direction west of Mellen, and farther north veers northeast, east of the Bad River. There for about 5 miles it runs between lava formations to the southeast and Lake Superior Lowland redbeds to the northwest.

Copper Falls State Park (more than 2,000 acres) west of State Highway 169 and north of Mellen, is situated where two rivers plunge over the Bad River Fault—the Bad River, which forms Copper Falls; and

Figure 3.11 Copper Falls, north of Mellen in Ashland County, where the Bad River plunges over the Bad River Fault. The rock is Keweenawan basaltic lava. (Wisconsin Department of Natural Resources. Photographer, Eugene Sanborn)

Figure 3.12 Tyler's Fork Cascades flowing over lava at Copper Falls State Park, upstream from Copper Falls and Brownstone Falls. (Wisconsin Department of Natural Resources. Photographer, Tom Grygo)

its tributary Tyler's Fork, where it joins the Bad River. The Bad River at Copper Falls drops 29 feet (9 m) over basaltic lava into a gorge. Tyler's Fork also tumbles over the lava, but in a series of falls and rapids totaling 70 feet (21 m) (figs. 3.2, 3.11, 3.12). Copper Falls is named for the coppery, or tea-like, tint of its water. Water acquires this coloration when it stays a long time in bogs having a certain kind of vegetation. This falls

area has been a popular place for a long time, and many old Indian arrowheads and pieces of worked copper have been found there.

After crossing the fault the Bad River flows first in a 100-foot-deep gorge walled by Keweenawan sandstone and conglomerate, then in a valley of red clay. It flows north through the center of the Bad River Indian Reservation to the marshy sloughs of northern Ashland County, where it enters Lake Superior.

Exploration goes on in the Keweenawan Area for bodies of copper ore large and rich enough for commercial development, and for other mineral deposits. It is spurred by the memory of the rich native copper mined from similar rock formations of Michigan's Keweenaw Peninsula.

Recently there has been speculation that oil and gas may exist in Keweenawan rocks, and interested companies have been exploring and leasing land in northwestern Wisconsin.

The Gogebic Range Area

The Gogebic Range Area has been one of the more intensively studied parts of the Precambrian shield in Wisconsin (fig. 3.1). That is because of the iron ore that was mined there in years past, and the ore still there.

The *Gogebic Range* is partly in Wisconsin and partly in Michigan. It is a narrow southwest-northeast belt of deformed and intruded Middle Precambrian sedimentary rocks. The entire range is 80 miles (130 km) long, extending from Lake Namekagon (pronounced Nah-muh-*cog*-un) in southeastern Bayfield County northeast to Lake Gogebic in Michigan. It is 3 miles (5 km) wide at most, and is slightly curved, with irregular bends at both ends (plate 1, fig. 3.2).

The western 53 miles (85 km) of the range are in Wisconsin and constitute the Gogebic Range Area. This Area's northwest border is the edge of the Keweenawan lavas. Its southeast border is the base of the Palms Formation, or of the underlying Bad River Dolomite where it exists. On the west the Gogebic Range Area reaches to the vicinity of Lake Namekagon, and on the east, to the Montreal River, beyond which the range continues into Michigan. Hurley and its twin city, Ironwood in Michigan, are on either side of the river where it cuts through the range.

The forming of the Gogebic Range was tied in with the sagging of the Lake Superior syncline, which parallels the range on the northwest. The sedimentary strata of the Gogebic Range, including its iron-formations, had been deposited in Middle Precambrian time. They became somewhat metamorphosed and tilted during the Penokean Orogeny, and the area was eroded. During the Keweenawan episode, when the syncline was deepening, strata of the Gogebic Area tilted to the northwest along with the southeast limb of the syncline. They kept tilting until they stood at a fairly steep angle (fig. 3.13). In places strata tilted to angles as great as 75 degrees, and even to a nearly vertical position. Sections of rock were displaced along faults that are parallel to, or at right angles to, the trend of the rock beds. Intrusions took place mainly in Late Precambrian time.

The iron of the Gogebic Range is in the Ironwood Iron-Formation, named for the city of Ironwood, Michigan (fig. 3.13). That formation, composed of interbedded iron-rich and silica-rich layers, is about 650 feet (200 m) thick. It overlies the Palms Formation (named for the old Palms mine), which is 450 to 550 feet thick. The Palms Formation consists of quartzite, slate and conglomerate strata, with the quartzite being the most resistant member (fig. 3.14). The Palms is conformable with the Ironwood; they merge with no

erosion interval between them. The Ironwood Iron-Formation and Palms Quartzite together form a ridge whose beveled summit rises above less resistant rocks on either flank. The Palms, being older than the Ironwood, lies along the range's southeast side. The Palms overlies the Bad River Dolomite where it is present, but where the dolomite is absent the Palms unconformably overlies Lower Precambrian granite, metamorphosed basalt, and other igneous rock of the area to the southeast.

Most of the Gogebic Range's iron ore has come from its Michigan section. In its Wisconsin section the Gogebic Range has produced more iron ore than any other part of the state—more than 71 million tons between 1884 and 1965, when mining ceased there. Wisconsin's two leading mines in this range, the Montreal and Cary mines, were in Iron County west of Hurley. Both began operation in 1886. The Montreal mine was the larger. As its rich ores were removed the mine shafts pierced farther and farther underground, and it became one of the deepest iron mines in the world at that time—4,335 vertical feet (1,320 m) deep. It produced some 46 million tons of ore, more than any other Wisconsin iron mine. After intermittent shutdowns it was abandoned in 1962. The Cary Mine was worked to a depth of 3,350 feet (1,020 m), and produced 17 million tons of ore. It was Wisconsin's last underground iron mine of that period to close, in 1965. The high-grade deposits that were economically within reach had been removed. Considerable quantities of accessible low-grade ore remain to be used whenever profitable methods of concentrating it for shipment are developed and the demand is great enough.

When geologists speak of an iron-bearing "range" such as the Gogebic they mean a whole belt of tilted strata (the iron-bearing ones as well as those not bearing iron). Within that belt, or range, there can be low as well as high features—valleys, hills or plains. So to geologists "the Gogebic Range" includes the complete belt of metamorphosed sedimentary rocks regardless of the topography, or surface configuration. It is in that geologic sense that the Gogebic Range is discussed here. However, to the lay person a "range" is a visual line of mountains or hills—in this instance, hills; and in popular usage "the Gogebic Range" means the line of hills that one sees. The Wisconsin section of that topographic feature is commonly called the *Penokee Range*.

The Penokee Range of Wisconsin is a slender band of hills generally not more than a mile wide (plate 4). To the east it is broader, a rounded swell, but to the west it is higher with summits of more than 1,800 feet. As far west as Mineral Lake in westcentral Ashland County it appears on the skyline as a wooded, uneven ridge. West of that it drops off, though its rock formations continue uninterrupted to Lake Namekagon. The range can be seen from Lake Superior.

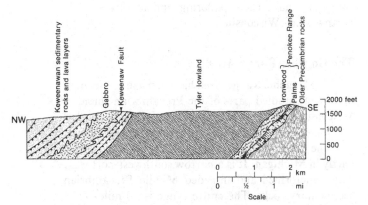

Figure 3.13. A cross-section through the Gogebic Range Area about 5 miles east of Mellen. It runs NW/SE, perpendicular to the trend of the range, from the Keweenawan Area on the left to older Precambrian rocks on the right. From northwest (left) to southeast the diagram depicts Keweenawan sedimentary rocks and lava layers, intruded by gabbro which formed an upland; the Tyler formation lowland; the Penokee Range (highest part of the Gogebic Range); and older Precambrian rocks. (After H. R. Aldrich, *The Geology of the Gogebic Iron Range of Wisconsin*, Bulletin 71, Wisconsin Geological and Natural History Survey)

Figure 3.14 The steeply tilted Palms Formation of quartzite and slate in Penokee Gap of the Penokee (Gogebic) Range. (From the historical files of the Wisconsin Geological and Natural History Survey)

The Penokee Range's highest summit is *Mount Whittlesey*—1,872 feet (570 m)—southeast of Mellen, just east of Penokee Gap (fig. 3.2). It was named for Colonel Charles W. Whittlesey, who studied the Wisconsin part of the range in 1849 and years following. It is said that he, by unclear handwriting, unintentionally gave the name Penokee to the range. Reportedly he named it the "Pewabic," which means "iron" in the Chippewa language, but a typesetter read it as "Penokie" and the name appeared that way in print. It became accepted and is in general use, while "Gogebic" is favored by geologists and mapmakers for the whole geologic range in both Michigan and Wisconsin.

The northwest side of the Penokee Range has greater relief, greater visual prominence, than its southeast side. On the northwest the range rises 100 to 600 feet above a parallel lowland about 2 miles wide. This narrow lowland was eroded in the relatively weak Tyler (or Tyler Slate) Formation, which consists of slate or argillite with graywackes and siltstones (fig. 3.13). The formation has a maximum thickness of about 10,000 feet. It overlies the Ironwood and was tilted to the northwest along with it. A railroad line and State Highway 77 were built along this Tyler lowland from Hurley to Mellen. Ore used to be shipped by rail along this route from mines near Hurley and Montreal to Mellen and on to the port of Ashland, where it was transferred to lake freighters. One of the giant shiploading docks that handled the ore still stands at Ashland.

Streams cross the Penokee Range through gaps, so it is not a continuous ridge but a series of long, elliptical hills lying end-to-end. Some gaps no longer contain streams. Much of the gap-cutting was done by ancient streams that must have been flowing across the region even while the Penokee Range was buried under younger rocks. (That old Precambrian range may have been covered by sediments of a Late Precambrian or Early Paleozoic sea.) As the covering of younger rocks wore off, the down-cutting streams would have reached the buried, more resistant range and begun incising themselves into it while the enveloping softer rocks were being removed. Down-cutting may have been facilitated by uplifting of the land. Undoubtedly some of the eroding streams encountered preexisting valleys, occupied them, and continued the excavation of gaps there. However, some gaps were freshly begun at that time. Faulting weakened the rock in places and allowed faster erosion there, as at the gaps of the Bad and Potato rivers. All gaps were deepened gradually as the range was exhumed and the surrounding surface of younger rock was lowered and ultimately removed.

The Bad River cut the lowest gap through the Penokee Range—Penokee Gap (fig. 3.15). Mellen, situated northeast of the gap on the Bad River (and in the Tyler lowland), is a highway and railroad junction. To

Figure 3.15 A view of Penokee Gap in the Penokee (Gogebic) Range about 1915 from a high point west of Foster Junction, looking southeast. The Bad River flows through the gap and a Soo railroad line was built through it. (From the historical files of the Wisconsin Geological and Natural History Survey)

the east, along the state boundary, the Montreal River cut a broad, shallow valley through the range, and its main west branch cut a gorge at which a dam was built, creating Gile Flowage.

Along the northwest side of the Tyler lowland is the *Gabbro Range* of the Keweenawan Area, parallel to the lowland and the Penokee Range. The Gabbro Range is almost as high as the Penokee but less noticeable on the landscape because when seen from the north its crest is lost against the backdrop of the Penokee Range (plates 3 and 4; figs 3.13 and 3.16).

The Northcentral Area

The Northcentral Area[2] lies between the Gogebic Range Area on the west and the Florence Area on the east, covering mainly Vilas County, much of Iron County, southern Ashland County, and parts of Sawyer, Price, Oneida, and Forest counties. Most of it lies southwest of the straight-line part of the Wisconsin-Michigan boundary (fig. 3.1 and plate 1).

The Northcentral Area is one of several parts of Wisconsin's Precambrian shield that early interested geologists, and so, geologically, it was investigated and at least generally mapped while many other parts remained essentially unexplored. Though the region is covered with drift 25 to 235 feet deep, there are outcrops; and drilling and subsurface detection instruments provided information about the buried bedrock.

Much of the bedrock dates from the Early Precambrian. During the Middle Precambrian, deformation created lowlands, or basins, where great amounts of eroded sediments accumulated, ultimately to become various metamorphosed sedimentary rocks. Volcanic materials collected there also. Folding of the

2. Dutton and Bradley's Butternut-Conover Area.

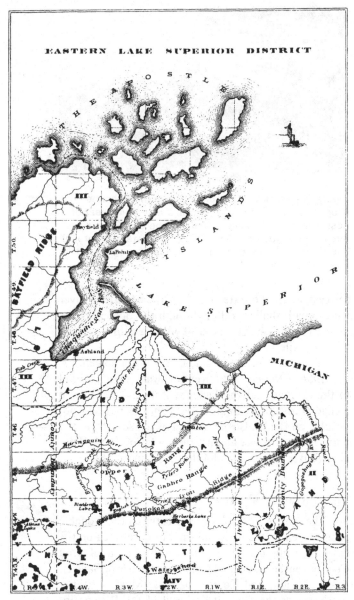

EASTERN LAKE SUPERIOR DISTRICT

Figure 3.16 The region of the Penokee, or Gogebic, Range as interpreted by R. D. Irving, published in 1880. (From T. C. Chamberlin, *Geology of Wisconsin*, vol. III)

ranges are not high, visible, topographic features, though they may once have been; they are ranges in the geologic sense of being belts of steeply tilted, iron-bearing rock.

Current studies in this and related parts of Precambrian Wisconsin are leading to new interpretations and greater knowledge of the region's geology and formative history.

The Florence Area

The Florence Area as delimited here is most of Florence County and the northern tip of Marinette County just to the east of it (fig. 3.1 and plate 1). It contains the Wisconsin part of the iron-bearing Menominee Range, or Menominee district, which extends east and northwest into Michigan (fig. 3.17). Most of the range is in Michigan, and that section of the range has produced more iron ore than has the Wisconsin section. In this case, too, the term "range" is used in the geologic sense and does not mean a landform feature.

Trending northwest-southeast across the Florence Area is a thick, Middle Precambrian sequence of metamorphosed sedimentary rocks interbedded with basalt, which have been complexly folded and faulted. The bedrock is drift-covered but commonly exposed on

rocks, much faulting, and many intrusions have interrupted the continuity of the preexisting rocks.

Iron-formation has been found in many locations throughout the region. This Area was explored for high-grade iron ore in the early 1900s, and more recently several of its districts were explored for lower grade ore that might be concentrated to a richer, marketable product, as well as for other minerals. Within the Area several possible mineral-producing districts were located, named from west to east, the Marenisco, Turtle, and Manitowish ranges (which trend southwest-northeast, generally parallel to the Gogebic Range), and still farther east, the Vieux Desert and Conover districts, where bedrock shows an east-west orientation. The Marenisco and Turtle ranges extend into Michigan. The

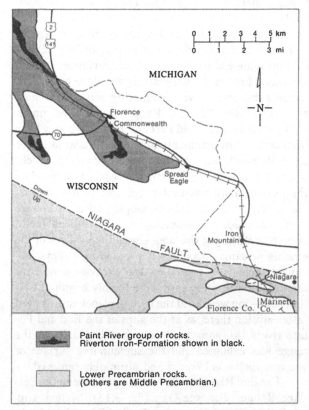

Figure 3.17 Simplified map of the Menominee Range region of Florence County. (After Carl Dutton)

ridges and along streams. The trend of the rock structure in Wisconsin determined the direction of the railroad and of U.S. highways 2 and 141, which run through Florence and Spread Eagle. The metamorphosed sedimentary rocks of the Menominee Range are separated from older igneous rocks along their southwest side by the *Niagara Fault* (fig. 3.17). It extends from eastern Vilas county across the northern tip of Forest County to the Brule River, and then southeast across Florence County, by the town of Niagara, into Michigan.

The Wisconsin and Michigan sections of the Menominee Range are not continuous. They differ geologically in age, kind of rock, and structure. Yet they came to be spoken of as one range because both are complex, deformed regions, aligned in the same direction, and because in the past ore from both was being sent to Escanaba, Michigan, for shipment to blast furnaces in Chicago, Pennsylvania and elsewhere.

Mining began in 1880 in the Menominee Range of Wisconsin. In 1882 T. C. Chamberlin wrote this about the Menominee mining district: "A region that five years ago was an almost unbroken wilderness is now dotted with thriving towns, and resounds with the scarcely interrupted rumble of passing trains."[3] The ore came from open-pit and underground mines near the towns of Florence and Commonwealth. Mining continued, though with interruptions, until 1937. More than 7 million tons of ore were produced during that time. Small shipments were made from 1952 to 1960.

The iron ore was taken from the Riverton Iron-Formation, which is generally about 500 or 600 feet thick, and is in the Paint River Group of rocks of Middle Precambrian (Animikie) Age.

The Wolf River Batholith Area

The Wolf River Batholith in northeastern Wisconsin is an exceptionally large, complex body of intrusions, as described in chapter 2. It extends from McCaslin Mountain (southeastern Forest County) on the north, and the vicinity of Crivitz on the northeast, to Waupaca on the south, and nearly to Wausau on the west. On the east it is overlain by Paleozoic rock (fig. 3.1 and 2.11).

The batholith is a composite of various igneous rocks that were intruded into older rocks during a relatively quiet period, early in Late Precambrian time. The many intrusions that comprise this batholith, though diverse, show a similarity of composition and age throughout. They are mainly coarse-grained granitic types, formed about 1.5 billion years ago.

Within the batholith region is an irregular area in western Shawano County that has attracted attention for some time.[4] It contains a number of outcrops and residual boulders that are moderately to highly mag-

netic. The possible value of the magnetic material has not yet been determined.

Certain areas within the batholith region may possibly contain uranium mineralization, but the potential is still not fully known.

Two features, not a part of the batholith but on the edge of the Area, should be mentioned.

McCaslin Mountain at the north end of the batholith is an elongated quartzite monadnock, probably of Middle Precambrian Age. A nearly east-west ridge, it extends from extreme southeastern Forest County into Marinette County, where it reaches its greatest height, 1,650 feet (503 m). It has a relief of 400 to 500 feet above the surrounding drift surface (plates 3 and 4).

Southeast of McCaslin Mountain, on the border of Marinette and Oconto counties, is *Thunder Mountain,* whose elevation is 1,410 feet. It is also composed of quartzite. Indians believed thunderbirds built their nests on its top, and stories were inspired by it. Though called a mountain, like all "mountains" in Wisconsin it does not have enough relief to fit that category and is more accurately described as an outstanding hill (plates 3 and 4).

The Barron Area

The Barron Area is that region on the western edge of the exposed shield underlain mainly by Barron Quartzite (fig. 3.1, plates 1, 3 and 4).

From eastern Barron County through northwestern Rusk and into southwestern Sawyer counties, quartzite hills and overlying drift form a rolling upland more than 25 miles (40 km) long from southwest to northeast, and up to 10 miles across. Hilliness lessens toward the north. Despite a heavy drift cover, the quartzite is apparent in many outcrops and loose-rock slopes and as angular boulders in the drift. The upland's highest point, west of Meteor (Sawyer County), is 1,801 feet (549 m) above sea level. West of the quartzite-cored upland are Lake Chetac (elevation 1,230 feet) and Red Cedar Lake (1,184 feet), and the towns of Rice Lake and Birchwood. Along the east side of the Area flows the Chippewa River (fig. 3.2). Northward the Barron Quartzite continues through a flatter, drift-covered area to Lac Court Oreilles and Grindstone Lake and beyond, to about five miles southeast of Hayward.

Unlike the monadnocks previously described, which have been completely disinterred from Paleozoic sedimentary rocks that formerly buried them, this larger Barron Quartzite feature is only partly uncovered. The

3. T. C. Chamberlin, *Geology of Wisconsin. Survey of 1873–1879* (Madison: Commissioners of Public Printing, 1883), vol. 1, p. 84.

4. Dutton and Bradley's Tigerton Area.

Barron Quartzite shown on bedrock geology maps, and described here, is only that part from which the overlying sedimentary rock has been eroded. The body of quartzite continues farther west, still overlain by Cambrian sandstone except for small scattered exposures. Between the Barron Area and the Keweenawan Area lava formations there are a number of Barron Quartzite outcrops. Some, southwest of Trego, are at the very edge of the Keweenawan Area. Most outcrops consist merely of residual broken blocks of rock at the surface. It is not known where and how far Barron Quartzite extends under Paleozoic rock. How it relates to Precambrian formations around it is not fully understood.

The Barron Quartzite upland constitutes a divide with streams flowing from it. Numerous gorges that contain no streams, or streams too small to have cut them, indicate that larger streams used to flow through this region before glaciers rearranged the drainage. A lobe of the last ice sheet, moving southwest, was retarded in crossing this hilly area and so apparently did not reach its southwest corner. Earlier, however, ice had overridden that corner too (plates 2 and 4). The melting ice front lay across the old hills for many years, dropping glacial debris there. Much of the upland's hilliness is caused by the unevenly deposited drift.

The most picturesque part of the Barron Quartzite upland is the southwest corner, which was not glaciated in the last ice advance. It is the west end of the *Blue Hills* (fig. 3.2), a range of hills that extends through the southern part of the Barron Area for about twenty miles, from eastern Barron County northeast into Rusk County, almost to the Chippewa River. The hazy-blue outline of these hills can be seen from the highways skirting them—U.S. highways 53 and 8, and State highways 40 and 48. The western end of the Blue Hills is high and abrupt. From it one can see the city of Rice Lake and look over the countryside for miles. The hills there are rough and rocky. In the canyon cut by Rock Creek quartzite is exposed in sheer walls. At the base of cliffs is deep talus (a sloping accumulation of fallen rocks), some blocks being as large as houses.

Typical Barron Quartzite is extremely hard and resistant to erosion, and difficult to break. When broken it is fairly smooth-surfaced and its edges are sharp, often knife-sharp. It is pink and red, and grades to purplish pink, white, gray, or sandy yellow. Some has dark red or pink iron stains in narrow, parallel straight or curved patterns. The sand grains of which this rock is composed are almost wholly quartz, variable in size but generally fine-grained. The cement binding them is silica. The sand of this now-metamorphosed rock was originally deposited on a broad, flat surface in shallow water. Some ripple marks are preserved, and crossbedding can be seen. The individual layered beds are thin, in most cases not more than a few inches thick; and the total thickness of the quartzite is about 600 feet (180 m).

Conglomerates, sandstones and shales are found in association with the Barron Quartzite. An especially sought-for soft rock occurs in small quantities, particularly in the Blue Hills. It is called *pipestone* because Indians made ceremonial pipes from it. Where it outcrops they quarried it from shallow pits, a number of which were in the hills east of Rice Lake (fig. 3.18). Pipestone originated from clay, and here is impregnated with iron which gives it a dark red color. In a moist condition when freshly dug, pipestone is easily cut and carved, but after exposure to air it dries and becomes hard. Because it is found in only a few scattered places, Indians came to this source from hundreds of miles away. They quarried it also at a site in southwestern Minnesota, now preserved as Pipestone National Monument. (That pipestone is named *catlinite,* after George Catlin [1796–1872], famous painter of Indian scenes.)

About 20 miles southeast of the Barron Quartzite upland, astride the middle of the boundary of Rusk and Chippewa counties, is a body of quartzite smaller in exposed area than the Barron but similar to it. This is the Flambeau Quartzite, which remains covered with Cambrian rock on all sides but the east (figs. 3.1 and 3.2, and plates 3 and 4). Its highest part, just south of the boundary in Chippewa County, is *Flambeau Ridge,* locally called *Flambeau Mountain.* It stands south of the eastward bend of the Chippewa River, which detours around it. (The Flambeau River joins the Chippewa north of the ridge's west end.) This slender, east-west monadnock is about 3.5 miles (5.6 km) long. Its south slope has a relief of 300 feet (90 m), and its north slope rises about 500 feet above the Chippewa River. Its highest elevation is more than 1,500 feet above sea level. At the east end of Flambeau Ridge lies a rounded quartzite hill (1,465 feet), and between this hill and the main ridge is a streamless valley that probably is a former channel of the Chippewa River. Flambeau Ridge is visible from a distance of more than ten miles, the best views being from the north and east.

Flambeau Quartzite is so like Barron Quartzite that it may actually be part of that formation. It also is mainly quartz grains well cemented with silica, is extremely hard, and breaks with fairly smooth surfaces and sharp edges. However, its bedding is more obscure, and it is better cemented and less broken. It is reddish brown to pale yellowish gray. The bedding in the main ridge dips north at a 75 to 85-degree angle.

The Wausau-Wisconsin Rapids Area

The Wausau-Wisconsin Rapids Area is on the south edge of the shield. It includes most of Marathon County (except the southeastern part) and parts of neighboring counties on the north, west and south (fig. 3.1). The larger cities of the Area are Wausau near the

a

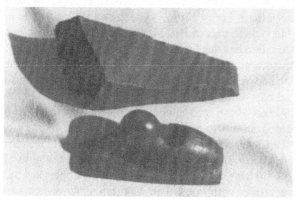

b

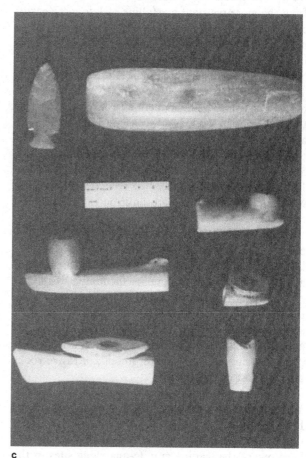

c

Figure 3.18 (a) A pipestone quarry in the Barron Hills east of Rice Lake. (Ralph V. Boyer) **(b)** Pieces of pipestone from near Rice Lake, before and after carving. This is a contemporary carving done by Bernard Ziegler. (Ralph V. Boyer) **(c)** Indian artifacts found in the Barron Hills, all from the same site. The arrowhead in the upper left is chert. The other items are pipes carved from pipestone. They appear light-colored but are really reddish brown. (By permission of the *Wisconsin Archaeologist* and John T. Penman, State Historical Society of Wisconsin)

center, Marshfield to the southwest, and Wisconsin Rapids[5] and Merrill near the south and north margins respectively, both on the Wisconsin River.

To the south and west of the Area, Cambrian rock still covers the shield. It is retreating along an irregular boundary, a ragged edging of sandstone "peninsulas" and "islands" (plate 1).

More of the Precambrian shield is exposed in the Wausau-Wisconsin Rapids Area than in any other part of the state. The last of the Pleistocene ice sheets stopped short of here, so this Area did not receive the thick accumulations and hills of drift that the Precambrian surface just to the east and north did (plate 2). Earlier in the Pleistocene ice had overrun this Area, but the drift that was deposited then is thin and not continuous. It is now so old, so long-weathered and long-eroded that in places it is even difficult to recognize as drift. Often little remains of it except its largest, most durable components—boulders.

Crystalline bedrock is generally near the surface in the Wausau-Wisconsin Rapids Area, just a few feet at most beneath the soil, old drift, and weathered rock. It is exposed in many places, especially along the channels of the Wisconsin River and its tributaries. Away from valleys the outcrops are smaller and scattered. Main valleys are floored with glacial material flushed south as the last ice sheet melted.

The Wausau-Wisconsin Rapids Area is drained by the Wisconsin River flowing south off the northern dome. The river enters the Area just north of Merrill, and from there to Wausau it meanders across the Precambrian surface in a valley of varying width and relief. At places north of Wausau the valley is 200 feet deep. It gradually becomes shallower toward the south. Dams across the river have created artificial lakes, such as Mosinee Flowage and Lake Du Bay.

5. Wisconsin Rapids is called Grand Rapids on older maps. Confusion with Grand Rapids, Michigan, led to the name change in 1920.

57

South of Nekoosa the river leaves the Area, passing from Precambrian to Paleozoic rock. Where it flows from the more resistant crystalline rock to more easily eroded sedimentary rock, falls and rapids formed. At many such waterpower sites along the edge of the shield lumber and grain mills were established early, and then other industries as well, and there towns were built. Today this section of the Wisconsin River Valley is the most industrially and commercially developed part of the Precambrian shield in Wisconsin. Although the falls and rapids were a major reason for the development of this region, they did keep the river from being navigable. Nekoosa was called "end of the rapids" by the Indians, and it was the head of river travel from the Mississippi until dams were built on the lower Wisconsin River.

Current field research is helping to clarify the geology of the Wausau-Wisconsin Rapids Area and relate it to the pattern of a broader region. Field work and also mineral prospecting are facilitated by the accessibility of the shield rocks. Known rocks indicate the presence of a volcanic-sedimentary sequence of rocks similar to others that have yielded commercially valuable materials. While various extrusive and metamorphosed sedimentary rocks are present, intrusive rocks are the most common.

Granitic rocks are widely distributed through the Wausau-Wisconsin Rapids Area, which is famous for high-quality granites used mainly for monument stone. Granite Heights, about 10 miles north of Wausau, has been one of the leading granite-quarrying sites in the state, producing stone of a variety of colors from gray to ruby red. It has been operating since 1895. West of Wausau and Mosinee disintegrated granite is dug easily with power shovels, and is used mainly for surfacing roads and also for decorating the ground in landscaping. Rhyolite, used for roofing granules, has been quarried northwest of Wausau and occurs in other places as well.

A fairly level plain across crystalline rocks is typical of much of the terrain. The main indentation is the Wisconsin River Valley. Rising above the plain are a number of old monadnocks, unyielding remainders of a once-higher Precambrian surface. In the Area's southern part are younger hills of another kind too, detached outliers of Cambrian sandstone.

It was long believed that *Rib Mountain,* the large monadnock southwest of Wausau, was Wisconsin's highest point (fig. 3.19). Its elevation is 1,924 feet (586 m) above sea level. Now it is known that Timm's Hill (1,952 feet) and Pearson Hill (1,951 feet) in southeastern Price County, and Sugarbush Hill (1,938 feet) in southern Forest County, which owe their altitude to glacial drift, exceed Rib Mountain in height. Still, Rib Mountain has the distinction of being the state's highest bedrock summit. It is an impressive hill because the Rib and Wisconsin rivers flow along its base on its north and east sides, and it looms high above them—about 760 feet above impounded Lake Wausau. Rib Mountain is an east-west ridge, slightly curved like a rib, about 3 miles long and more than a mile wide. It is the site of *Rib Mountain State Park.* The lower flanks of this hill have a gradual rise, whereas the upper part has steep sides with talus. The northern slope is most abrupt. Rib Mountain is composed of some of the most resistant rock in nature—extremely hard, firmly cemented, massive (homogeneous and unbroken) quartzite, which here is generally translucent and off-

Figure 3.19 Winter view of the north side of Rib Mountain. White snow-covered ski runs contrast with the wooded slope. (Harold O. Krueger)

white in color. Ripple marks and cross-bedding show that originally it was sand deposited offshore in an ancient sea. Folding has caused its once-horizonal beds to be tipped to a nearly vertical position.

Mosinee Hill and *Hardwood Hill* are smaller monadnocks in the Wausau vicinity, also composed of highly resistant, remarkably pure quartzite—white to pale pink (because of iron staining).

Mosinee Hill (sometimes called Mosinee Hills) has two summits, which give it a dumbbell shape. These summits are one and two miles south of the east end of Rib Mountain, on the west bank of the Wisconsin River west of Rothschild. Upper Mosinee Hill, the larger one to the north, is 1,605 feet (489 m) above sea level; Lower Mosinee Hill, 1,470 feet. Both have gentle slopes on their west side and are steeper on the east side facing the river.

Hardwood Hill (or Hardwood Ridge), also quartzite, is about 3 miles southwest of Rib Mountain. It covers half of a square mile, is dome-shaped, and reaches an elevation of about 1,600 feet, which is about 400 feet above the nearby valleys.

Powers Bluff, another well-known monadnock, is near Arpin in northcentral Wood County, about 35 miles southwest of Rib Mountain. It is generally called fine-grained quartzite, but may be recrystallized chert. Many slowly decaying crags of this rock crown the bluff, whose summit elevation is 1,481 feet. It rises about 300 feet above the surrounding area, standing among remnant hills of thin Cambrian sandstone; and still remaining on its base are patches of Cambrian conglomerate, deposited hundreds of millions of years ago when the bluff was a wave-dashed island in the Paleozoic sea.

The Black River Falls-Neillsville Area

The Black River Falls-Neillsville Area is an offshoot of the exposed shield trending southwest across Clark and Jackson counties. Its branching, antlerlike shape is mainly that of the course of the Black River and its tributaries where they have cut through Paleozoic sedimentary rocks, revealing the underlying Precambrian crystalline rocks. Neillsville is near the center of this elongated Area, and the city of Black River Falls is near its southwestern end. The exposed rocks, many of which are gneisses, granites and schists, tell of intrusive and metamorphic events, relating to the shield's development, dating back to Archean time.

Drift cover is light or absent. The Clark County part of the Area is in the region of thin, old drift; the southwestern part in Jackson County is in the Driftless Area. (Fig. 3.1 and plates 1 and 2.)

Near Black River Falls several small, metamorphosed sedimentary Precambrian knobs containing low-grade iron-formation protrude upward through Paleo-

zoic strata. Mining was attempted there in 1857, but it was not successful then because the iron content of the material was low and rich ores were obtainable elsewhere. Later, when the rich ores were in shorter supply, this low-grade ore (taconite) had a market. In 1969 open-pit mining of iron ore resumed at Iron Mound, one of the knobs about 6 miles east of Black River Falls. At the mine, before the ore was shipped to blast furnaces, it was put through a process that eliminated much worthless rock. The mined ore, which is only 30 to 35 percent iron, was concentrated into pellets that were about 60 percent iron (fig. 2.10). These were shipped by rail to iron-and-steel plants in East Chicago, Indiana. Mining at Iron Mound was interrupted in 1982.

The Central Shield Area

The part of Wisconsin's Precambrian region not yet described in this chapter is here treated as a unit and called the Central Shield Area (fig. 3.1). This broad region—probably the least-understood sizable part of Precambrian Wisconsin—should not be considered a region of uniformness, but rather one of diversity and complexity; much of its boundary is loosely drawn. This Area incorporates most of the part of Precambrian Wisconsin that until recently was left blank on geologic maps, or was just vaguely portrayed.

This Central Shield Area, which lies in the southern part of the broad zone that was involved in the Penokean Orogeny, has an extremely complex rock structure, showing the effects of much faulting and folding and widespread volcanic activity. A high proportion of its rocks are volcanic, while to the north more rocks are of sedimentary origin. Blocks of rock were slowly uplifted, tilted or dropped, and other processes of deformation and intrusion greatly disturbed the terrain. Erosion of higher landforms provided sediments which were deposited in lower places. All became metamorphosed during millions of years of geologic change. Long erosion wore the irregularities down to a fairly level surface.

Over the years it has been a frustrating task for limited numbers of field workers to map this expanse of bedrock where outcrops are scarce, vegetation is dense, and drift is up to 200 or 300 feet thick. How much more challenging to construct a three-dimensional picture of the bedrock. The rocks, some of which date from Early Precambrian, have been altered through a number of major tectonic events. Until recently it was virtually impossible to successfully explore the formative history of the region, but now stepped-up field work and current scientific techniques and insights have been bringing newly discovered facts into focus. Geologists are able to speculate more intelligently about how this region was involved in the amalgamation of the Precambrian shield.

Among the intrusive rocks of the Central Shield Area granites are common, and some of them are of outstanding quality for monument and building stone. Near Amberg in central Marinette County several quarries in granitic rock have produced high-quality stone. At one time this was one of the most important granite-quarrying sites in Wisconsin, famous for the variety of colors and textures of its granites. Colors range from red through pink to gray, and some of the finest-grained and coarsest-grained granites have been found here. Nearby to the southwest, at Athelstane, an exceptionally coarse-grained gray granite has been quarried.

Mining companies have intensively explored certain districts in the Central Shield Area, and have found several ore deposits that are rich enough to mine.

One mile southwest of Ladysmith in Rusk County is a six-million-ton deposit that contains about 4 percent recoverable copper with minor amounts of zinc and traces of gold and silver. It is lens-shaped, about 50 feet wide, 2,400 feet long, and 800 feet deep.

A major zinc-copper deposit has been found 6 miles south of Crandon and north of Little Sand Lake in Forest County. The mineralized zone appears to be about 5,000 feet long, 1,500 feet deep, and 125 feet wide, but it may be larger. The ore averages about 5 percent zinc and 1 percent copper, with substantial lesser values of silver, gold, and lead. Containing about 80 million tons of ore, it ranks among the six largest known deposits of this kind in North America.

Southeast of Rhinelander in Oneida County, near the Pelican River, is a zinc-copper body that consists of more than 2 million tons of ore averaging nearly 5 percent zinc and 1 percent copper. It is about 1,000 feet long, 50 feet wide, and 650 feet deep.

The Buried Shield

The Precambrian shield concealed under Paleozoic rock remains largely unknown, but drilling and probing with geophysical instruments have provided some information regarding the types of rocks there, the location of faults, and the topography that was already sculptured on the shield's surface when the Cambrian sea rolled over it. Old river valleys and hills have been detected beneath the Paleozoic rocks. Some hills and ridges on the buried part of the shield have greater relief than many that are considered major features on the exposed part. There are even entire ranges of buried hills (fig. 1.4).

One large, completely buried range, named the Fond du Lac Range, is surmised to run from southwest of Lake Winnebago through central Fond du Lac County, into Calumet County east of the lake. The range seems to stand several hundred feet above the average level of the Precambrian plain. Along this zone

quartzite and slate have been reached at shallow depths, signifying that the hilltops are not far below the surface.

Another buried range, the Waterloo Range, extends from the towns of Portland and Waterloo (at the west end of the boundary between Dodge and Jefferson counties) northeast about 12 miles and perhaps beyond. This range was more easily discovered than the Fond du Lac because its highest parts are uncovered and show as ledges and rounded knobs (fig. 3.20). They have been scoured and reduced by glaciation, which left striations and furrows on them and carried boulders away from the parent masses. The boulders are strewn for miles to the southwest. The outcrops of the Waterloo Range are extremely hard, as might be expected—strongly metamorphosed quartzite, dark red to gray in color, with some metamorphosed conglomerates and thinly layered, foliated rocks present. Much more erosion of the covering rock must take place before the Waterloo Range fully emerges: drills go through more than 700 feet of sedimentary rock on either side of the range before reaching the Precambrian surface.

The "underground" Fond du Lac and Waterloo ranges may be just as extensive as the Baraboo Range; and perhaps other ranges, still unknown, exist beneath Wisconsin's sedimentary rocks.

Pieces of Precambrian rock found in Cambrian conglomerates that were deposited around a number of the shield monadnocks (including the Waterloo Range and some of those mentioned later) are evidence of past ocean burial. When the Cambrian sea flooded over the Precambrian lowlands the high landforms became rocky islands and reefs; their shores were beaten by

Figure 3.20 An outcrop of Waterloo Quartzite (a summit of the buried Waterloo Range) on a farm two miles northwest of Hubbleton in southwestern Dodge County on State Highway 19. Field notes of 1939, the date of the picture, record that because of the hard quartzite underlying this farm it was impossible for the owner to drill a well and drinking water had to be carried from a spring. (Milwaukee Public Museum. Photographer, Kenneth Vaillancourt)

waves, and pieces of rock broken from them were deposited in the conglomerate sediments accumulating around them.

For as long as a hundred million years or more, sea-floor deposits forming sedimentary strata built up around and over the Precambrian hills and ridges. These strata protected the hills from further erosion for up to a few hundred million years—longer in some areas than others—until the Paleozoic seas had gone and the sedimentary cover was eroded down far enough to re-expose their summits. The softer sedimentary rocks wore away faster than the resistant Precambrian rocks, and are still being eroded faster. Eventually (barring intervening deposition) Wisconsin's partially exhumed monadnocks will be entirely uncovered—still standing high—as many to the north, where sediments are thinner, already are: the Penokee Range, Rib Mountain, McCaslin Mountain, and others.

Precambrian Inliers

Exposures of Precambrian rock are seen within the Paleozoic rock area in scattered places beyond the margin of the uncovered shield (plate 1). Such exposures of old rock completely surrounded by younger rock are termed *inliers*. In southcentral Wisconsin Precambrian inliers are found as far south as Lake Mills (Jefferson County). The Waterloo "hilltop" outcrops described in the preceding section are Precambrian inliers (fig. 3.20).

The inliers are generally small; most of the main ones are only a fraction of a square mile in area. The largest is the Baraboo Hills, covering about 75 square miles (200 sq km). Typical inliers are rounded, ice-scoured mounds or knobs, and ledges; these are the tops of old hills and ridges again showing themselves where the softer Paleozoic rock was worn away and the glacial drift is thin. Not all of the inliers are high features, however. Some that have been exposed by the valley cutting of streams are lower than the general surface. The inliers are mainly quartzite, rhyolite or granite, and many contain intrusions. These "peepholes" of Precambrian rocks—once-molten, contorted, intruded, crystalline, or otherwise igneous or metamorphic in character—are a distinct contrast to the younger, nearly horizontal sedimentary strata framing them.

In and near the Upper Fox River Valley are a number of inliers (fig. 3.21). Northern Columbia County has several, of rhyolite mainly and other igneous rocks. There are a few outcrops—low, rounded, weatherworn hills and other exposures—in the township of Marcellon near the center of the county's northern boundary. Six miles (10 km) to the north in southern Marquette County is Observatory Hill, which is about 200 feet high (elevation 1,100 feet) with rhyolite ledges exposed at the top. Five miles west of

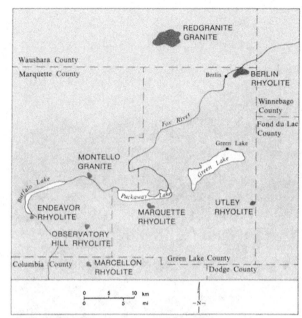

Figure 3.21 Some of the notable Precambrian inliers of the Upper Fox River Valley area. (After Eugene I. Smith)

Observatory Hill at the southern end of curved Buffalo Lake near Endeavor (sometimes called the Moundville area) are three rounded rhyolite outcrops, the largest being at the east end of a low bluff.

At the eastern end of Buffalo Lake in the city of Montello is an elliptical mound of granite, about a third of a mile long and about 40 feet high. The close-textured, weathered surface shows almost no tendency to decompose. Bright pink or reddish monument stone of exceptionally high quality has been quarried there (fig. 3.22). When General U. S. Grant died in 1885 a search was conducted to find the most durable attractive granite for polished stone coffins for him and his wife. The search covered three continents. The granite finally selected was from Montello. The coffins were cut from single granite blocks more than 10 feet in length, and are in Grant's Tomb in New York City.

Just south of Puckaway Lake in Green Lake County, southwest of Marquette, is a group of rhyolite hills. Twelve miles to the east near Green Lake County's eastern boundary is the Pine Bluff exposure at Utley. Pine Bluff—a 100-foot-high mass of bare, glacier-scratched rhyolite—rises out of the marshy floodplain of the Grand River. The resistance of this hill (representative of many other exhumed Precambrian monadnocks) is demonstrated by the fact that the Paleozoic rock around the hill was worn away by the river and glaciers while the hill remained. Quarrying of rhyolite began here in the early 1800s and continued for several decades. Rhyolite similar to that at Utley is found at Berlin, 17 miles to the north, where the exposure is in the shape of three elongated domes. An

Figure 3.22 Part of the granite quarry in the city of Montello which has supplied monument stone of exceptionally high quality and durability. (Gwen Schultz)

Figure 3.23 Science Hall on the University of Wisconsin–Madison campus. Its foundation is constructed of grayish metamorphosed rhyolite, an extremely resistant rock, from Berlin in northern Green Lake County. (The Archives, University of Wisconsin-Madison)

almost-black, dense, and durable stone quarried in the Berlin and Utley areas is commonly called granite, but it is rhyolite (fig. 3.23).

(Though many Precambrian exposures are classed as the same kind of rock—for example, as granite or rhyolite—they may vary considerably in texture, composition and structure, and in many instances are also associated with other crystalline rocks.)

Large exposures of rhyolite and granite occur west of Berlin along the boundary between Green Lake and Waushara counties, and north to around Spring Lake and Redgranite in southern Waushara County. A number of quarries are found through this area.

About 4 miles south of New London in southeastern Waupaca County (sometimes called the Mukwa area) is a granite exposure consisting of a row of three large and three small, elongated domelike outcrops, the highest rising nearly 70 feet.

Necedah Mound in Juneau County is a partly uncovered, half-mile-long hill of highly metamorphosed quartzite. Its roundness distinguishes it from bold younger hills of sandstone in that vicinity. The town of Necedah is built at its base. The mound has an elevation above sea level of 1,099 feet (335 m), and its highest

and most rocky side rises 170 feet (50 m) above Necedah's streets.

Other sizable quartzite outcrops are the Hamilton Mounds in northern Adams County, and a cluster of exposures in marshy southwestern Wood County—the main ones being North Mound, Middle Mound, and East Mound, all composed chiefly of a similar massive quartzite. North Mound, the largest of the three, lies about 3 miles northwest of Babcock; it is oval and partly talus-covered, and rises 200 feet above the flatlands. Middle Mound, a mile east-northeast of it, is low and irregular in shape and about 50 feet high. The smallest, East Mound, lies a quarter of a mile farther east.

Of the many other Precambrian inliers, some conspicuous ones are given names but others receive little attention. In this transitional Precambrian-to-Paleozoic zone mounds and crags of sedimentary rock, some of which have angular and jagged shapes, may be more noticed than the longer-eroded, rounded, or low-lying Precambrian outcrops.

The Baraboo Hills

The *Baraboo Hills* in southcentral Wisconsin are the state's largest, most renowned Precambrian inlier, extending from central Sauk County into westcentral Columbia County. (See plate 1, including cross-section, and plates 3 and 4.) Composed of quartzite, they are the high parts of a worn-down synclinal structure. They are still partly buried by horizontal Paleozoic strata, which are layered all across southern Wisconsin.

These hills are also called "The Baraboo Range," but since there are actually three (landform) ranges within the Baraboo Hills, that name is not used here. The name "Baraboo Bluffs" is sometimes used, especially for the Devil's Lake area.

Geologists have been greatly interested in the Baraboo Hills since their significance was first realized in the mid-1800s, for they encapsulate a record of important Precambrian, Paleozoic and Ice Age events in a compact setting. The region interested geologists also because it contains an iron-bearing formation, which was once mined even though its iron content is low.

The ranges and hills that make up the Baraboo Hills together form the shape of an arrowhead whose tip points a little north of east. The "arrowhead" is about 25 miles (40 km) long and has a maximum width of about 10 miles in its middle and western section. Its outline is formed by the prominent South Range and the less obvious North Range, which are the upturned limbs of a syncline. Their east ends join a few miles southwest of Portage (just south of the Baraboo River near where it enters the Wisconsin). The tip of the arrowhead, where the ranges meet, stands out as a high bold point overlooking the rivers below (figs. 3.24 and 5.28).

The higher parts of the Hills rise more than 700 feet (200 m) above the valley of the Wisconsin River. If all Paleozoic rock and drift were removed from around their base and the Hills appeared as they did in Late Precambrian time before burial, they would be much more imposing. With relief in places increased by about 500 feet, some parts would stand about twice as high above the ancient plain.

The quartzite of which the Baraboo Hills are largely composed overlies rocks that existed in this area before the long period of Precambrian sedimentation. The basement rock is poorly exposed. Rhyolite appears in several places below the quartzite along both the North and South ranges. The largest exposure is on the north side of the North Range at the gap (the Lower Narrows) where the Baraboo River leaves the synclinal lowland, mainly west of the gap (figs. 3.24, 3.25). The rhyolite there has yielded dates that suggest it is between 1.8 and 1.65 billion years old. Granite and diorite also are exposed in a few places.

The Baraboo Quartzite originated from sand that was deposited upon the foundation of igneous rocks in a shallow-water environment early in Late Precambrian time (fig. 2.5). During many millions of years it accumulated in horizontal beds to a great depth—more than 4,000 feet (1,220 m)—while slowly the waters rose or the land subsided. The sand probably came from eroding highlands uplifted during the Penokean Orogeny. The beds' cross-stratification shows that currents bringing the sand flowed from north to south. The sand was mostly unmuddied, remarkably pure, medium-to-coarse-size quartz grains. Quartz in solution cemented the grains together, forming sandstone, and iron gave the rock its color. It is generally pink and ranges into maroon, purple, gray, and, in a few places, white. The rock shows stratification and, occasionally, ripple marks,

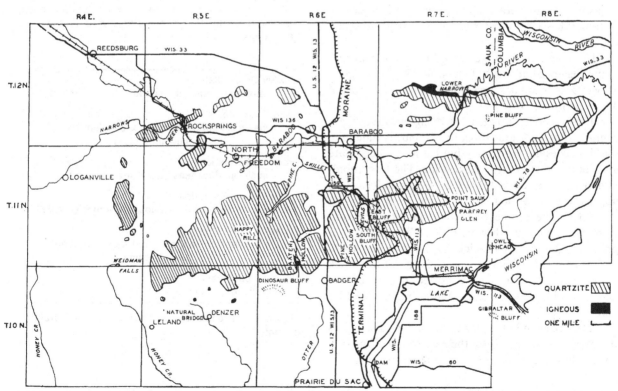

Figure 3.24 The Baraboo Hills area as mapped by F. T. Thwaites in 1958, showing major geological features. (The *Transactions* of the Wisconsin Academy of Sciences, Arts and Letters, vol. 47)

Figure 3.25 Members of a geography field trip examining an outcrop of rhyolite on the southeast side of the South Range of the Baraboo Hills in Columbia County. (Robert W. Finley)

Figure 3.26 These ripple marks in quartzite in the North Range of the Baraboo Hills were made in sand along a shore in Precambrian time. Later, when the sand had become rock, this bed among others was tilted to a vertical position. (Robert R. Polk)

and much of it has alternating light and dark layers (fig. 3.26).

While about 85 percent of the sediment deposited then was well-sorted sand, other materials were deposited too. Conglomerate is present below and interlayered with the sand, along with layers of once-clayey material ranging from a few inches to several feet thick.

Uplift and synclinal folding took place, and the Precambrian sandstones became metamorphosed to a massive, extremely hard quartzite; and the weaker interlayered conglomerates and clayey materials became metamorphosed, to greater or lesser degrees, to schist and slate.

Lying conformably above the Baraboo Quartzite is the Seeley Slate Formation, which is fine-grained, gray or green, and has a maximum known thickness of 370 feet (fig. 3.27).

The Freedom Formation, overlying the Seeley Slate conformably and gradationally, is mostly dolomite (hard magnesium limestone), but has interbedded slate and chert and is locally iron-bearing. The formation's minimum thickness is 1,000 feet. Dolomite forms the upper part and various iron-bearing rocks form the lower part. The iron-bearing unit is similar to the iron-formation found in other districts of Wisconsin. It is generally about 200 feet thick, but locally at least 400 to 500 feet. Ore bodies in it were 20 to 30 feet thick before being mined; they consisted mainly of red hematite with some limonite, and averaged about 53 percent iron.

The Freedom Formation is overlain in some places by two younger units inside the eastern end of the syncline. They are the Dake Quartzite and the Rowley

Rowley Creek Slate—
    maximum known thickness, 149 feet (45 m)

Dake Quartzite—
    maximum known thickness, 214 feet (65 m)
        (unconformity?)

Freedom Formation—
    minimum thickness, 1,000 feet (300 m)

Seeley Slate Formation—
    maximum known thickness, 370 feet (113 m)

Baraboo Quartzite—
    thickness more than 4,000 feet (1,220 m)
        (unconformity?)

Rhyolitic basement—
    thickness unknown

Figure 3.27 Precambrian stratigraphy of the Baraboo district.

Creek Slate. (Some geologists have questioned whether those units are additional formations. They were first known from descriptions of a quartzite and a slate in core samples obtained during subsurface drilling. Those samples resemble Baraboo Quartzite and Seeley Slate, and it was felt they may be parts of those formations rather than younger ones.)

The Baraboo Hills syncline formed asymmetrically. The beds of its north limb (seen in Ableman's Gorge north of Rock Springs) were tilted up at a steep angle, becoming vertical and even slightly overturned in places (fig. 3.28). The beds of its south limb (seen in Devil's Lake Gorge) were tilted more gently, dipping 20° to 40° northward into the trough (fig. 3.29). In the east the structure of the syncline is simple, the trough narrowing at the end like the front of a rowboat. The central and western parts of the syncline are wider and have additional, complicated folds.

Figure 3.28 In the walls of the Upper Narrows, north of Rock Springs in the North Range of the Baraboo Hills, can be seen quartzite beds that were originally horizontal layers of sand but have been vertically upturned. (Robert W. Finley)

Figure 3.29 Looking east toward East Bluff at Devil's Lake in the South Range of the Baraboo Hills. The pattern of tree growth helps to show how the quartzite beds dip gently to the north (left) into the synclinal trough. (Roger Peters)

Long erosion smoothed the general surface of the area and left the syncline's upturned, exposed quartzite beds as rounded hills and ranges. The syncline's northern limb protruding at the surface creates the North Range, and the southern limb creates the South Range. These ranges have summit elevations of about 1,000 feet (300 m) to 1,500 feet above sea level.

During Cambrian and Ordovician times sedimentary deposits slowly accumulated over the terrain. Even the highest features were completely buried and thus preserved. Residual sandstone and limestone from those Paleozoic periods are found atop the quartzite hills, evidence of their burial. Subsequently erosion re-exposed their upper parts. While the obdurate quartzite hills have suffered little reduction since being uncovered, the softer sedimentary rocks around them are continuously wearing away (fig. 3.30).

Paleozoic sedimentary rocks still overlie low places among the Baraboo Hills, and still partly fill the lowland between the North and South ranges, which extends along the down-folded trough—the axis of the syncline. This lowland, in which the city of Baraboo is located, has an elevation of 800 to 900 feet, and is drained by the Baraboo River (fig. 3.31).

Figure 3.30 On top of the East Bluff at Devil's Lake are many potholes such as these (which are dappled in shadows). These water-polished depressions range in size from a few inches to 3 feet in diameter and depth, resembling bowls or more deeply drilled holes. When stream action at that location was strong enough to scour these holes is not known. It may have been when Paleozoic rocks were being eroded from the buried bluff tops, or perhaps in the Pleistocene when glacial meltwater poured over this place if the ice reached this far (a supposition less generally accepted). (Ralph V. Boyer)

Figure 3.31 Aerial view of the dark, wooded Baraboo Hills and the lowland between, looking east from Baraboo. The North Range on the left and the higher, less broken South Range on the right converge in the distance. They are the limbs of a synclinal trough. The Baraboo River meanders across the foreground and on along the right side of the lowland. (Robert W. Finley)

Figure 3.32 Aerial view of Devil's Lake looking north into the synclinal lowland. Talus drapes the lower part of East Bluff at the right of the lake. A river used to flow through this gap before glacial deposits blocked both ends. (Madison Metropolitan School District. Photographer, Ron Austin)

Connecting the North and South ranges at the west end of the lowland is the shorter, interrupted West Range which extends north-south for about 7 miles. It lies a few miles east of State Highway 23. This quartzite range is about 2 miles wide, and its highest elevation is more than 1,400 feet (425 m). The Baraboo syncline continues west beyond the West Range, but there it is still covered by Paleozoic rock, which is gradually being eroded, and has about the same elevation as the exposed quartzite hilltops.

The eastern part of the Baraboo Hills region was glaciated—east of a line running roughly north-south through the middle of the North Range, the western outskirts of Baraboo, and the Devil's Lake gap area. West of that line the hills received no drift, but the central lowland there, though unglaciated, contained a glacial lake in Late Pleistocene time and is therefore covered with lake-bed sediments which overlie the Paleozoic rock.

The South Range is the most prominent part of the Baraboo Hills and constitutes at least half of the exposed quartzite area. It is a nearly continuous ridge with only one deep gap, *Devil's Lake Gorge* (fig. 3.32). Though the gorge was stream-cut, no stream flows through it now. It is blocked at both ends by glacial drift and contains the lake. The river that carved the gorge may once have cut across the North Range (perhaps at the Lower Narrows) and exited here. Much of the cutting was already done by the end of Precambrian time; there are still traces of unremoved Cambrian sandstone in the gorge. The gorge may have been deepened later. (It is described further in chapter 6 in regard to glaciation.)

The Baraboo River enters and leaves the central lowland through gaps in the North Range (fig. 3.24). It enters at the west end of the North Range through the Upper Narrows, or Ableman's Gorge, north of the town of Rock Springs (formerly called Ableman), and leaves at the east end of the range through the Lower Narrows.

The Upper Narrows is a curved, 200-foot-deep gorge barely wide enough to allow State Highway 136 and a railroad line through beside the river. Along the road in this gorge is Van Hise Rock, a stout pillar or monolith that plainly shows the quartzite's vertical tilting and other characteristics. It is named for Charles Van Hise, geologist and former president of the University of Wisconsin at Madison. At the gorge's north side are famous exposures of the unconformity between the Precambrian quartzite and Cambrian sandstone, with old shoreline conglomerates indicating where waves of the Cambrian sea broke against the ancient hills. Quarrying has altered the natural appearance of the gorge. A small part of it has been designated a State Scientific Area (figs. 3.33, 3.34).

The Baraboo Hills display some of the highest, most rugged "hard-rock" topography in the Midwest. People travel great distances to see this large exposure of tilted metamorphic rock which is strikingly different from the simpler, "softer," flat-lying sedimentary rock that predominates south of the Precambrian shield.

Baraboo Quartzite is some of the hardest rock in the world. It is not porous, and it weathers and erodes ever so slowly. Though greatly fractured, its loose fragments and talus remain sharp-edged even after countless years of exposure. In the span of a lifetime one would probably notice no change, no disintegration, in this rock, but of course disintegration is going on, aided by the scablike lichens growing on the rock, by plant roots prying into cracks, by the expansion of freezing water in openings, by the infrequent tumbling of talus blocks, by the beat of geologists' hammers and tread of hikers' shoes (figs. 3.35 and 3.36). Because of its durability, Baraboo Quartzite is used for such things as open-hearth furnace lining, railroad ballast, and abrasives. In the past it was used too as macadam aggregate for roads, and as paving blocks.

Iron-formation was discovered in 1887 in Freedom township, and was mined just west of La Rue for paint pigment from 1889 to 1899. Then in 1900 ore-grade iron was found there, and mining began a few years later. The ore was taken from the lower member of the Freedom Formation. The mines were the Cahoon Mine south of Baraboo and the Illinois and Sauk mines southwest of North Freedom. The Cahoon Mine operated from 1916 to 1925, and the Illinois Mine from 1904 to 1908, and in 1916. Each shipped a total of more than 300,000 tons of ore. Records are not available for the Sauk Mine, which opened in 1903. The total amount of ore shipped from this district may have been a million tons. Flooding of the mines was partly responsible for their closing.

Figure 3.33 The Upper Narrows (Ableman's Gorge) seen from the north in the year 1920. In the center foreground, between the railroad and highway that make use of this gap, stands Van Hise Rock. (From the historical files of the Wisconsin Geological and Natural History Survey)

Figure 3.34 At the Upper Narrows of the North Range of the Baraboo Hills one can see horizontal strata of Cambrian (Galesville) sandstone unconformably overlying the eroded surface of the Precambrian Baraboo Quartzite, the beds of which are vertical. This clearly shows that the Baraboo Hills were submerged in Cambrian time. Those sedimentary rocks are being eroded from the more resistant quartzite hills, as they already have been from the summits of the eastern part of the ranges. (Roger Peters)

Figure 3.35 Devil's Doorway, a rock formation high on the East Bluff overlooking Devil's Lake, displays characteristics of Baraboo Quartzite: the manner of jointing, the sharpness of breakage lines even after long weathering, the resistance to erosion, and color banding. (George Knudsen, Wisconsin Department of Natural Resources)

Figure 3.36 Along the trail leading up the south side of the East Bluff at Devil's Lake one sees how quartzite blocks fallen from the bluff's upper face settle as talus below. (George Knudsen, Wisconsin Department of Natural Resources)

The Baraboo Hills, with their gorges and glens, their scenic rock formations and other natural features, attract nature-lovers. The flora is of special interest too. Because of the variety of microenvironments within the Hills, an uncommonly large assortment of plants grows in this small area.

The Indians built hundreds of animal-shaped effigy and burial mounds in this area. Many have been removed by farming and road-building. A county park a few miles northeast of Baraboo was established to protect Man Mound, the only known extant Indian earthwork monument in human form. It is shaped like a gigantic human figure 214 feet (65 m) long from head to foot and 48 feet (15 m) wide at the shoulders.

A Broad View of the Quartzites

Many clues to the Precambrian are fragmentary and unconnected, but, as we have seen, when certain ones are properly related they give revelations of past events. Just so, several late Precambrian quartzites, widely separated but similar, are especially elucidating (fig. 3.37).

It was recognized long ago that the quartzite of the Baraboo Hills is almost identical to the Waterloo Quartzite which outcrops 30 miles (50 km) to the southeast. Both originated as thick accumulations of almost-pure quartz sand and contain thin layers of conglomerates and schistlike rocks. Both have similar coloration and are characterized by color banding, ripple marks, dull glassy luster, and extreme hardness. Their cross-bedding orientations are comparable. Radiometric dating indicates that both quartzites were deposited in the same general time period, early in Late

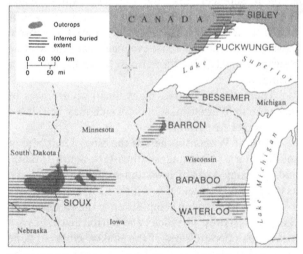

Figure 3.37 Major quartzites in and near Wisconsin, possibly related. (R. H. Dott, Jr., and I. W. D. Dalziel)

Precambrian (or possibly late Middle Precambrian) (fig. 2.5). Furthermore, the Waterloo Quartzite outcrops appear to be part of a buried syncline that pitches eastward along an axis in about the same direction as that of the Baraboo Hills syncline. These synclinal ranges possibly are monadnock remnants of the same formation that was originally continuous over at least a large part of southern Wisconsin and was involved in an episode of folding.

Other quartzite inliers in southcentral Wisconsin are similar in composition to the Baraboo and Waterloo rocks, supporting the idea of a former wide zone of deposition of quartzite-forming sands. As one looks farther afield this concept takes on even broader proportions.

Barron Quartzite to the northwest, about 600 feet (180 m) thick, is remarkably like the Baraboo and Waterloo quartzites. It is not surprising that people who know both the Baraboo Hills and the Blue Hills section of the Barron upland see a strong resemblance in the topography and natural appearance of those two regions. Did the Baraboo-type quartzite of southern Wisconsin extend far enough north to be continuous with the Barron Quartzite? Much of the Barron Quartzite is buried, as is much of the southern quartzite, and rock between the separated monadnocks has been eroded, so the present and past extent of these quartzites is not known.

Farther north, lava formations of the Lake Superior syncline rest on a basement of Lower Keweenawan quartzite which may be the continuation of a widespread quartzite formation. Bessemer Quartzite is exposed along the upturned southeast side of the Lake Superior syncline in Wisconsin and Michigan, outcropping in only a few places. Its maximum thickness is about 300 feet. Like the Baraboo, Waterloo and Barron quartzites, it is remarkably pure. In depositional sequence it is below (older than) the Keweenawan lava flows, and above (younger than) the Middle Precambrian Gogebic formations. It overlies the Tyler Formation.

There are, of course, many small occurrences of quartzite in Wisconsin, some of which may be related to the larger formations.

West of Wisconsin there is the Sioux Quartzite of southwestern Minnesota, southeastern South Dakota, and adjoining parts of Iowa and Nebraska. It may be as much as 3,000 feet thick in South Dakota. This rock is virtually identical in composition to the Baraboo and Barron quartzites. The Sioux and Barron quartzites are located on opposite sides of the sunken synclinal zone of Keweenawan lavas which extends south from the Lake Superior basin along the Wisconsin-Minnesota border. Both of these quartzites include beds of dark-red pipestone, and both lie unconformably upon deformed and metamorphosed older rocks, as does the Baraboo Quartzite. Dates obtained from rocks underlying the Baraboo and Sioux quartzites indicate they may be equivalent in age.

The distribution of similar thick beds of quartzites raises the thought that for millions of years sand from the eroding shield may have been collecting in a fairly uniform shore or basin environment that extended from eastern Wisconsin at least as far west as South Dakota. The depositional zone may have been even broader.

Along the northwest shore of Lake Superior, in northeast Minnesota and adjacent Ontario, are the Sibley and Puckwunge formations, which contain quartzites that may have formed while the quartzites to the south were forming. They bear an uncertain but suspected relationship to the Barron Quartzite in that they, too, lie alongside the Lake Superior syncline and have a similar structure and composition. They may be equivalent to the Barron, and perhaps other quartzites.

Because of similarities among these various quartzites, some geologists believe that several or many of them may be remnants of one deep, vast blanket of sandstone. The sands would have been laid down in a large body of water on the south side of the shield, or across a low part of it, during a stable period following the Penokean Orogeny. All of the quartzites have pink or red coloring. All are compositionally and structurally alike, and similarly cross-bedded. All were deposited in like manner in comparable environments—on a slowly subsiding, shallow-water shore resembling a continental shelf. Most are known to lie unconformably on older igneous and metamorphic rocks.

In order for sand to accumulate to thicknesses of hundreds of feet over many millions of years, the crust below must have been stable. But later it was disturbed and the thick sandstone blanket (changing to quartzite) was rumpled. These layered rocks, especially those near the shield margin, buckled into folds, creating a rough terrain which was subdued by long erosion.

Looking Ahead

As we have surveyed the different regions of Wisconsin's Precambrian shield we have repeatedly been reminded of one fact: most of the shield is covered and unseen. Even so, its importance cannot be over-estimated. It provides stability, and its rocks hold mineral resources.

If the ice sheets had not come and deposited drift over these ancient crystalline rocks, they would be at the surface in more than a third of the state. They would be far better known, and there would be considerably more mining and quarrying activity.

Surely Wisconsin's Precambrian rocks are a storehouse of undiscovered useful minerals—some commonly utilitarian, some rare and of great value. Undoubtedly new deposits will be located in the near future because exploration is intensifying; but other deposits will not be known for centuries or millenia, if the human race endures that long. In a way, it may be good that thick glacial drift blankets most of this area, hiding the contents below, saving them from immediate use and wastage, for what our civilization obtains easily it tends to squander. Those treasures were formed in the distant past under special conditions and in Earth's slow-working subterranean forge. When they are mined they cannot be recreated. They cannot just be made when wanted, nor can they be reused indefinitely.

It is predicted that sometime in the future when an ever-expanding population has depleted the more obvious deposits, Wisconsin's long-secluded reserves will be unearthed. Then they will probably be needed and appreciated much more than they would be now. Until such time they lie undisturbed awhile longer, under forests and lakes, farms and recreation land, and restful glacial hills.

The whole geologic base of the shield area of Wisconsin's northland promises to be increasingly valuable in years to come, not only its rock and mineral resources, but its waters, its natural scenery, its crop and forest lands. The people will say which assets they deem most important. Some favor the development of mining and ore processing industry. Others would curtail development to safeguard the prime vacation land and undefaced landscape; clean rivers, lakes and air; and quite, serene way of life. Here tourism is a leading industry, and here healthfulness and beauty of the outdoor environment have been zealously protected. Can the recreational, natural environment and mining industries coexist satisfactorily? At the writing of this book, the future of northern Wisconsin's shield area is a matter of great concern and controversy.

4

The Paleozoic Story—Layer Upon Layer

The story of the Paleozoic Era in Wisconsin is largely a story of the sea, of many seas. It tells of how they came and went several times, and of the deposits laid on their floors while they were here. Now hundreds of millions of years later those marine deposits, turned to rock and located far from the withdrawn sea, strongly influence the activities and well-being of all who live upon them.

Chapter 1 described how the fluctuating Paleozoic seas deposited variegated layers of sedimentary materials upon and around the Precambrian shield. The fact that those sedimentary layers were variegated is all-important. Because they are of different types, they give diversity to the surface where they are exposed, and also to conditions underground.

To us today, commonplace happenings of the Paleozoic Era may seem inconsequential—like those that determined what earth materials and biological remains settled from the sea waters or were left on the beaches . . . what localized concentrations of substances occurred . . . what chemical reactions . . . what fluctuations of water depth. Yet those remote events—even the minor changes—determined the composition and sequence of the dissimilar rocks that formed at different times and places. Also, percolating groundwater slowly modified the rock by bringing and carrying away certain substances, and cementing some materials more strongly than others. (Of course, such events pertaining to the formation of sedimentary rocks were not limited to Paleozoic time. Sedimentary rocks of Precambrian time, discussed in preceding chapters, were likewise affected, as are those forming today.)

Bedrock has important effects on physical and cultural aspects of our lives. That can be borne in mind in a general way as we look at the characteristics of Paleozoic sedimentary rocks in this chapter. In the following chapter more specific and regional examples of these effects will be presented.

The Importance of Differences Among Rocks

Some rocks exposed at the surface are quite resistant to erosion; others are less so. Their relative strengths have helped determine what is lowland or highland, and therefore where streams and rivers run, and where roads, railroads, cities, and farms are situated. As rocks disintegrate, the materials they release help determine how fertile and arable the resulting soil will be, what natural vegetation will grow on it, what crops will be successful, and generally how the land may be used and developed. Also, some rocks can serve as building stone, whereas others lack the durability or attractiveness for that purpose. Some rocks can be crushed and processed for industrial, agricultural and construction uses. Some rocks are composed of clay materials of a particular quality needed to make brick, tile and other ceramic products. Rocks that have concentrations of usable mineral ores attract mining and related industry. Limestone that has been dissolved by groundwater may contain caves which draw tourists and give earth-scientists avenues for investigating subterranean conditions. Sometimes just the appearance and configuration of rock-controlled terrain affect the land's desirability and ownership value, especially for home building and recreational development. Resistant rock ridges have turned glaciers this way or that, and so have further influenced the character, uses and value of the land.

Even inaccessible rocks far below the surface or hidden under glacial drift have an important bearing upon our lives. This is particularly true in regard to our supply of water, a vital natural resource. Much of the water we use is drawn from bedrock some distance underground; and the nature of the rock from which the water comes is a major determinant of its volume and quality and the reliability of supply. Rivers and lakes, from which we take water too, are filled not only by surface runoff but also by seepage from natural reservoirs in bedrock and unconsolidated deposits, including glacial drift.

Sandstone has pore spaces among the grains through which groundwater can slowly filter.

Shale is compact and greatly retards or prevents the movement of groundwater through it.

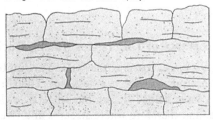
Limestone and dolomite develop openings and cracks through which water can move rapidly.

Figure 4.1 Characteristics of sandstone, shale and limestone.

Many crystalline Precambrian rocks are so compact that they contain very little groundwater, but thick strata of porous sedimentary rocks can hold large amounts. It is estimated that in Wisconsin nearly half of the water for industrial, municipal and private use comes from Paleozoic sedimentary rocks.

The several kinds of sedimentary rock differ markedly in their ability to accept and store the water that falls as rain and snow; to make and keep that water clean; and to allow its movement to other regions (fig. 4.1).

Sandstone, which has pore spaces and good filtering capability, is generally the most reliable, steadiest supplier of high-quality water to wells and springs.

Shale can absorb considerable amounts of water, but because it is finer-grained and more compact than sandstone it is largely impermeable; that is, it does not let water pass through easily. But by this very impermeability shale acts as a sealing layer, helping to concentrate and direct groundwater flow in adjacent, more permeable strata.

Limestone and dolomite generally have much less storage capacity. Because they can be dissolved by acidic groundwater and therefore tend to have openings along bedding planes and fractures, they characteristically contain many interconnecting cracks and caverns through which water can move freely and rapidly. Pollutants from the surface that are washed into such openings can enter the groundwater directly, and water from those rocks may be unsafe for domestic use without considerable purification. Of course, water in any kind of rock is subject to pollution, but the chance of pollution is greater in limestones and dolomites.

Groundwater picks up minerals from the rocks it travels through, and the presence of certain minerals makes water less desirable for household, agricultural and industrial purposes. Limestone and dolomite add calcium and magnesium to groundwater, and are chiefly responsible for making water "hard." Sandstones cemented with calcareous material also yield hard water. Hardness causes water to be unusable or objectionable for some industries, and makes it bothersome to use in the home. If a water-softener is needed, that is added expense. The hardness or softness of your household water affects you personally as you bathe, wash your hair, or do laundry. In hard water soap does not lather and leaves a scum. Also, hard water leaves limy precipitates in water pipes, tanks and teakettles which must be removed periodically. If groundwater picks up too much iron it has an objectionable taste and smell, and it leaves rusty stains on laundered articles and porcelain plumbing fixtures, and troublesome deposits in pipes. "Mineral water" is considered healthful by many people. A number of wells and springs in Wisconsin supply such water, known for its natural purity and medicinal effect. (The health claims made for some waters have been questionable or unfounded.)

So we see that the composition of the various sedimentary rock layers, even those unseen and untouched, are important to us. Industrial, agricultural and recreational enterprises, and the general economy of an area, as well as the personal lives of its people, are greatly affected by the rocks below, by the volume and quality of water from those rocks, and by the soils, mineral resources, and other earth materials derived from them. It is therefore understandable why geologists use every available means to learn the composition of rocks at and below the surface, to ascertain the thickness and horizontal continuity of various strata, and to trace the routes along which water moves through them.

Overall View of the Sedimentary Strata

In some areas of the world the thickness and structure of the sedimentary rocks are dramatically evident—for example, where they are steeply tilted or severely faulted and form mountains, or where deep canyons cut into them in barren plateau regions, exposing great cross-sections of rock. In the Wisconsin area the Paleozoic sedimentary rocks are less impressively displayed, for they experienced only small measures of tilting and bending and little faulting, and much of the surface is covered with drift; but the effects these unobtrusive rocks have upon local geography and people's daily lives are just as basic and important as those of the more dramatically exposed rocks elsewhere.

In eastern Wisconsin the Paleozoic sedimentary rock layers dip gently downward toward Lower Michigan, where they form a circular basin of strata that

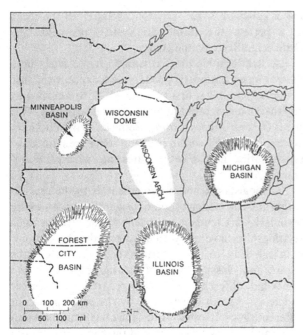

Figure 4.2 Sedimentary rock basins around Wisconsin.

are warped downward toward the center, resembling a stacked set of saucers. The strata of eastern Wisconsin thus form part of the western edge of this basin (fig. 4.2 and plate 1 cross-section). Another such basin exists in Illinois, and the strata of southern Wisconsin dip down in that direction. The lower Paleozoic strata are slightly upbent in the Wisconsin Arch, the gentle fold that extends lengthwise from central Wisconsin into northern Illinois. Its crest, which runs north-south, crosses the Baraboo Hills region and passes just west of Madison. To the west and southwest the sedimentary strata slant down off the Wisconsin Dome and Wisconsin Arch into southeastern Minnesota and Iowa. The Mississippi River has cut a gash into these rock layers, exposing matching beds on both sides of its valley.

Learning How the Layers Formed

Geologists are aided in visualizing the original deposition of Paleozoic sediments in Wisconsin and neighbor states by the fact that the rocks have been little disturbed, and also by observing similar deposition now going on in shallow-water coastal areas. One such area is the Gulf of Mexico shoreline zone from Florida to Texas. During the Paleozoic in the Wisconsin area, as now in the Gulf region, the shoreline was irregular with bays, lagoons and deltas. Sediments disgorged by many rivers and eroded from the coast were weighed, sorted and deposited by waves and longshore currents.

Wind too played a part in the sorting and abrading of rock particles and in transporting the finer ones. Its role must have been especially important before land plants had become established, lessening its effect.

Even if a shoreline environment had remained static, there would have been differences in the kind and amount of deposits accumulating in one place and another, but the differences would be multiplied if the sea retreated and advanced, which is what did happen many times. So the rock layers vary in dimension and composition. Layers that formed under uniform conditions are identified individually as "units."

Some units are less than an inch thick, while others are hundreds of feet thick. Some are just local, while others extend in overlapping, coalescing sheets from Wisconsin through several states or into Canada. Some merge vertically with their overlying neighbor, changing gradually or abruptly from one kind of rock to another. For instance, a sandstone may become progressively less sandy and more shaly in successive overlying beds, until it becomes a true shale—a gradational change that commonly indicates increasing distance from the shoreline. By studying the transitions in deposition of sediments geologists can tell how ancient marine environments were modified with the passage of time.

Geologists can chronicle past events also by studying sites where neighboring units do *not* merge vertically, where an unconformity separates strata (fig. 1.11). If there is such a break in continuity, represented by an eroded surface, the geologist sees the following situation in retrospect. The sea withdrew and erosion took over. As erosion progressed the terrain came to resemble an emerged coastal plain like that today along the Gulf of Mexico in the southern United States. The surface rock layers were cut into by streams, creating low hills and ridges with valleys or wide lowlands between. The strata were worn away more in some places than others, depending upon the location and cutting power of streams, the relative resistance of the rocks, the location of faults, and so on. For countless years surface rock weathered and was dissected by gullying and shifting streams. The rock layers that were removed altogether are lost from the record. Then the sea returned and began depositing a new, relatively smooth floor upon this irregularly eroded surface. First the sea penetrated only the valleys, so the deposits were not continuous horizontally because of the intervening hills and rises, but as the water level rose the sedimentary floor covered wider areas.

The sea withdrew and advanced many times in Wisconsin as shown by sequences of deposition in rock beds and by unconformities. Because of changing conditions of deposition, the stratigraphic layering is a patchy arrangement of assorted sedimentary units—some small or thin, others quite broad or thick. Rock units may merge with overlying and underlying ones, or be separated by an eroded surface. Edges of the units taper out and interfinger with others. That is why it is not easy to figure out their arrangement and relationship.

Cross-section exposures, as in cliffs and bare slopes, show the vertical sequence of strata at those sites. Drill holes reveal the subsurface sequence in many other places, but provide far from uniform coverage. In Wisconsin it is required that whenever a high-capacity well (such as a municipal, industrial or irrigation well) is drilled, representative samples of the cuttings, or rock chips, brought up by the drilling machine must be sent to the Wisconsin Geological and Natural History Survey to be analyzed and recorded. Cuttings must be sent in also from smaller, low-capacity wells drilled for the state or federal governments, as in parks and waysides.

Core-drilling machines, which bore deep into rock to bring up cylindrical cores intact, are utilized where the bedrock is being carefully explored for a specific purpose. Wide-angle-lens cameras are lowered with lights into some of the larger holes to obtain pictures of the rock walls. Electrical impulses sent into the ground also tell something of the character of the rock.

For areas where rock data are lacking, interpolation must suffice. At a given place a certain vertical sequence of strata may be known. Some distance away one may find a somewhat altered sequence; a rock unit may be thinner or missing, or a new unit may be interposed. As rock researchers study the findings from many such separate sites, and as they map the extent and relationship of various layered rocks, they fill in the stratigraphy of the unseen, unprobed areas as best they can, according to their judgment. Thus a broad picture of a region's bedrock stratigraphy is drawn. In time, details of this picture will become more fully known (fig. 1.23).

Terms Used to Classify Sedimentary Strata

So that researchers anywhere can trace, correlate and conveniently refer to specific rock units in separated sites, it is necessary to have a mutually understood classification with terms defined. (See plate 5.)

A *formation* is a distinct mappable rock unit, one that is fairly homogeneous or has distinctive characteristics, and is a main body of rock in the local classification.

Formations are subdivided, where feasible, into thinner or local units called *members,* which have their own special characteristics. For instance, a shale formation may be composed of several members, one of which may be sandier than the overlying or underlying one, and others which may contain certain ingredients or fossils not found in the adjoining members. Members are subdivided into *submembers* where distinctions can be made. Two or more formations may be classified together as a *group.*

The terms so far defined apply to rock stratigraphic units. These rock units are related to time units called *series,* which are combined into larger categories called *systems.* A system includes all the rocks of a period: the Cambrian System of rocks, the Ordovician System, and so on.

A stratigraphic unit is named after a site where there is an exceptionally good, typical exposure of that unit, or where it was first recognized. The name is usually that of a geographical feature or a nearby city. This site is the unit's *type location.* Some units in Wisconsin are named for Wisconsin sites and some are named for type-location sites in other states into which the unit extends.

Geologists are not in agreement as to the classification and correlation of some rock units in cases where boundaries are inadequately defined and stratigraphic relationships have not yet been worked out. And new data can lead to new interpretations. Therefore the names and classifications of stratigraphic units are somewhat different in charts prepared by different people, or even in charts done by the same person at different times. In other words, the classification is tentative and being worked out, and the names are subject to change.

In older geologic writings and diagrams we see rock-unit names that are no longer in use. And we see names that have been shifted to other categories: a name that used to be applied to a formation may now be applied to a group, and other switches of names among subdivisions have occurred. Also, some names of rock units became names of time units, and vice versa. So one should be watchful when referring to older publications and maps, and know how the old unit names compare with current ones. Some of the significant name changes will be pointed out as units are described here.

Although rock sequences are observable in only a limited number of places, a general pattern is apparent as the sequences from various sites are matched. In this pattern geologists look for cycles of deposition.

Cycles of Deposition

A *cycle,* in reference to sedimentary rocks, is a sequence of several main rock types that recurs, with variations. It records a series of geologic events and physical conditions that was repeated in the same order a number of times.

In each cycle of events the sea transgressed, or advanced over the land, and then regressed, or withdrew, leaving as evidence of its presence certain kinds of sediments which ultimately became rocks. Then while the land was above water, the surface rocks were carved by erosion. When the sea moved in again another cycle of deposition began upon that irregular eroded surface. In this way cycles recurred repeatedly.

Some cycles are plainly read in rocks, whereas some are indistinct and questionable. Opinions may differ as to how many cycles occurred in a given area because

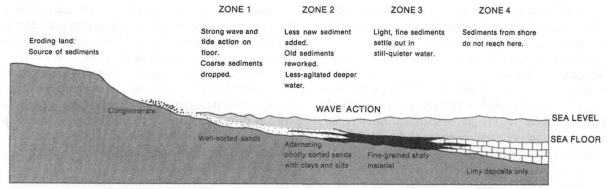

ZONE 1   ZONE 2   ZONE 3   ZONE 4

Eroding land:
Source of sediments

Strong wave and tide action on floor. Coarse sediments dropped.

Less new sediment added. Old sediments reworked. Less-agitated deeper water.

Light, fine sediments settle out in still-quieter water.

Sediments from shore do not reach here.

WAVE ACTION

SEA LEVEL

SEA FLOOR

Conglomerate

Well-sorted sands

Alternating poorly sorted sands with clays and silts

Fine-grained shaly material

Limy deposits only

Figure 4.3 Transitional zones of deposition at a given time on a shallow sea floor.

Physical Characteristics

Type 1: Sandstone of predominantly medium-grained, well-sorted, clean sand. (May include some wind-deposited sand.) Thick-bedded to medium-bedded. In places conglomerate is at the base.

Type 2: Sandstone of poorly sorted sands tending to be more coarse-grained than Type 1. Some clay and silt. May have been reworked by burrowing organisms or water action. Thin-bedded to medium-bedded. Transitional. Alternating reworked sands and muds.

Type 3: Shale or shaly sandstone.

Type 4: Limestone (which changed to dolomite)

Zone of Deposition

1. Formed in shallow marine environments, such as beaches or nearshore locations, where wave and current action was strong.

2. Formed in water quieter than the Type 1 environment where there was mainly reworking of bottom sediment, with little or no new deposition. Commonly seaward of Type 1 on the shallow-water shelf.

3. Formed on the shallow-water shelf in water quieter than Type 1 and 2 environments, where fine-grained particles could settle out. Farther from shore than Type 1, and usually also farther from shore than Type 2.

4. Formed where sands and silts were not present and where material produced by marine organisms accumulated. Seaward of the other types, or in clear shallow waters as in lagoons or on tidal flats.

Figure 4.4 Main rock types in cycles of sedimentation (pertinent to Wisconsin).

some rock units have been entirely removed by erosion, and at some levels in the strata it is not clear whether or not an unconformity (erosion interval) divides layers, and, if it does, whether it represents a full or partial withdrawal of the sea.

Cycles of deposition logged in the hard "pages" of rock tell by "chapters" what was happening on the regional scene. In the Wisconsin area three main transgressions, or advances, of the sea stand out in the Late Cambrian, and at least two in the Ordovician. (There may have been lesser advances as well in those times.) Other cycles followed in the Silurian and Devonian. The sea may have returned to Wisconsin in the Mississippian and Pennsylvanian periods, but rocks from those periods are not found here to supply proof (fig. 1.9).

Of the several systems of Paleozoic rock in Wisconsin, the Cambrian and Ordovician cover the widest areas and afford the broadest opportunity for studying cycles (plate 1). They are best seen in the driftless southwestern part of the state. In the east, from Door County south, they are overlain by Silurian rocks, and, in a small lakeshore strip, by Devonian rocks as well.

Throughout the glaciated area drift forms an additional cover, but bedrock does show through in many places, and drillings give scattered samples of what lies below (plate 2).

The main types of sedimentary rock—sandstone, shale, limestone (predominantly dolomite in Wisconsin) and conglomerate—have variations in composition, as described in chapter 1. There are more mixed types than pure types. It is the order of deposition of these rock beds, as well as their gradation from one to another, that provides the basis for identifying cycles of sea transgression and regression.[1]

These sedimentary rocks were formed mainly from the deposits laid down along the shore and on the offshore shallow-water shelf. In regard to cycles, these rocks can be classified into four main types, which formed in four different environments, or zones of deposition, as shown in figures 4.3 and 4.4. (Relatively clean sand was deposited first in each cycle and the resulting sandstone is therefore called Type 1.)

1. The depositional sequence of four main rock types and the outline of cycles in this chapter follow a classification used by M. E. Ostrom. (See plate 5 and Ostrom 1964 in Selected References.)

All of these environments existed simultaneously in Wisconsin in Paleozoic time (as they do today in other regions), shifting their locations as seashores migrated, and changing as the kind and amount of available material changed.

With rise of sea level relative to the land, the depositional zones migrated shoreward. Conversely, with a drop in sea level they migrated seaward. Just a minor sea level change over shallow-water shelves could lead to broad shifts of these zones (fig. 4.5, 4.6).

One common way in which the sediments forming the main rock types might have been deposited can be imagined by picturing the sea moving into the Wisconsin area very slowly. In a given location the first, and lowest, significant layer deposited is sand, which forms such features as beaches, bars and dunes. Then the shoreline migrates inland, laying down its floor of sand as it does so. Back at the original shoreline location, the upper part of the deposited sand is being reworked by water movement and seafloor animals, and

Material eroded from the land is carried into the sea where waves and currents sort and deposit it:

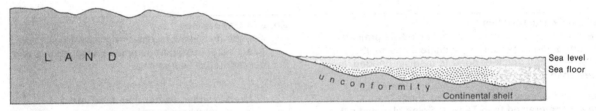

The sea slowly rises (or land sinks) and waters encroach upon the land:

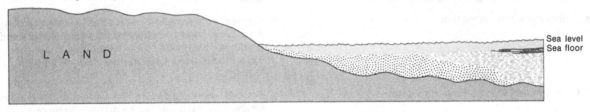

The sea continues to rise and transgresses farther inland:

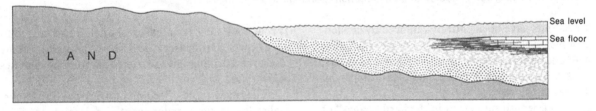

The sea spreads still farther inland:

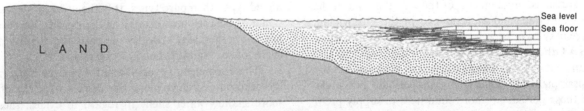

Note the succession of sediments upward at any given point--from sand, to mixed sand and finer particles, to all fine particles, to limy material. No scale is intended.

Figure 4.5 A sequence of sedimentation left by a shallow sea transgressing without interruption upon a continental shelf.

Figure 4.6 Two scenes along the Lake Michigan shore at Kohler-Andrae State Park south of Sheboygan. One **(a)** is a beach at a time of low water level. The other **(b)** is about a half mile north at a time of higher water level, where waves have been eroding sand dunes. The comparison suggests how an encroaching sea might have advanced inland in Paleozoic time. (George Knudsen, Wisconsin Department of Natural Resources)

mixed with finer material present in the less agitated or somewhat deeper waters. As the shoreline moves even farther inland, water over the original zone becomes deeper and the floor is less disturbed by the breaking of waves and the flow of currents. In this quieter environment fine silty and clayey sediments can settle out; the mud thus deposited will later become shaly rock. All this time the shoreline moves continually inland, until eventually sand and silt from the shore no longer reach the original location. There waters are now quite still, and only limy precipitates and the remains of sea life accumulate on the floor, to become limestone and dolomite.

Then the sea starts to recede, and each zone of deposition recedes with it, shifting seaward. Some of the previously formed deposits on the sea bottom become exposed and are partly or completely removed by

erosion. Eroded materials are carried by streams to the retreating sea.

A perfect, complete depositional cycle like the one described is hard to find in nature. No two cycles are alike—in duration, environment or available materials. Cycles are not rhythmically periodic, as the term may imply. Many are only partly completed before the sea changes direction, moving seaward when it had been moving inland, or vice versa. At times the sea floor itself slowly rises or falls while sea level is fluctuating, upsetting the cycle further. Throughout geologic time and throughout the world, fluctuations in sea level could have resulted from many causes, such as changes in the shape of ocean basins, and crustal warping that raised or lowered the land in some places, and large-scale glaciation. (When thick continental ice sheets held a large quantity of water frozen on land, the sea level fell worldwide; as the ice melted, sea level rose.)

In looking for signs of cycles in rocks, it is helpful to see the first layer, the clean well-sorted sandstone, lying upon an uneven, eroded surface. This shows that the land had previously stood above the sea and that erosion had taken place before the arrival of the sea with its sandy shoreline. In other words, there is an unconformity between the cycle's first sandstone layer and the underlying rock. Unconformities are distinguishable in some sedimentary sequences, but not in others.

Transitional beds may exist between unlike sedimentary strata, recording the gradual change from one to the other, but elsewhere the contrast may be sharp and the change abrupt. Then it may be hard to tell whether an unconformity exists there or whether deposition was continuous. The picture is complicated too if the sea did not withdraw completely, but went through minor fluctuations or partial cycles, creating irregularities in the pattern.

Deposits left during the sea's withdrawal were not as well preserved as those that were covered over while the sea moved inland. As the sea withdrew, much of the deposited material that emerged was reworked, or eroded and redeposited near or beyond the retreating shoreline.

The preceding explanations should show why there are many problems in outlining cycles. Because geologists are still trying to time the ins and outs of the Paleozoic seas, a full, detailed account cannot be given here, but some of the known events are described.

A Cycle's Four Main Sedimentary Rock Types

This outline summarizes information about sedimentation cycles and associated rock types in preparation for the descriptions of specific Paleozoic strata that follow (fig. 4.4).

The four main types of rock in sedimentary cycles, with special reference to the Wisconsin area, are these:

Type 1, which was deposited first, is the lowest layer in a cycle. It is thick-bedded to medium-bedded, relatively clean, fairly uniform sandstone, consisting of rounded quartz sand grains of mainly medium size. Its sands, having been sorted well, are essentially free of fine clays and silts. Lime materials too are absent. Because of vigorous, destructive wave and current action along the shore, fossils are few. Locally there may be conglomerate at the base of the sandstone, including rounded pebbles, cobbles and boulders that the first inland-moving waves were abrading and reducing in size (figs. 4.7, 4.8). Cross-bedding and ripple marks, which are indicators of the direction of water movement, show that the sediment in the Wisconsin area was washed by currents moving from north and northeast to south and southwest, corresponding to the trend of the ancient

Figure 4.7 A beach near the north boundary of Racine County on Lake Michigan illustrates on a small scale the conditions that exist at some places along the shore of a transgressing sea. Waves erode the land (especially during storms) and work over the eroded material, leaving well-sorted sands at the shore. Cobbles and boulders at the foot of the eroding cliff (too heavy to be washed into the water) remain. If the sea advanced over them they would form conglomerate at the base of sand laid above them. (Wisconsin Department of Natural Resources, Scientific Area Preservation Council)

a

b

Figure 4.8 (a) Parfrey's Glen, one of several picturesque glens eroded in Cambrian sandstone on the flanks of the Baraboo Hills. It is on the south side of the South Range, about 4 miles east of Devil's Lake. Deep in the glen, where icicles remain well into spring, the author surveys the rock walls. (Robert W. Finley) **(b)** Close-up of a wall of Parfrey's Glen showing conglomeritic sandstone. The rounded rocks of the conglomerate are quartzite—pieces eroded from the Baraboo Hills when they were islands in the Cambrian sea. Waves at the shore rolled and dashed the fallen pieces and rounded them by abrasion. As sea level slowly rose, these shore deposits were covered by other marine sediments which buried the hills completely. Now erosion (exemplified in the work of the small stream that is carving this gorge) has uncovered the tops of the resistant quartzite hills and continues to remove sedimentary strata from around their base. At times huge blocks of rock fall into the glen. Figure 1.6 gives a closer look at the conglomerate. (Ralph V. Boyer)

shoreline around the southern curve of the submerging shield. Rock units (discussed later) that fit this type are the Mount Simon, Galesville, Jordan, New Richmond, and St. Peter sandstones (plate 5). These names are given also to cycles they initiate; for instance, the Mount Simon Cycle began with the deposition of Mount Simon Sandstone.

Type 2 is the reworked sandstone unit that overlies Type 1 and was located seaward of it. It is transitional between the Type 1 and Type 3 units and usually shares characteristics of each. The texture of its sand grains is coarser than that of Type 1 because washing has removed many of the finer grains. Its contact with Types 1 and 3 may be either sharp or gradational. This rock type is composed of interbedded layers of well-sorted quartz sandstones and poorly sorted silty and clayey sandstones, along with some shales, sandy shales, and conglomerates. Some limy material may be present. The alternating of quite different beds reflects shifting environmental conditions and variations in wave and current energy. Fossils are more common here than in the more agitated zone of Type 1, and burrows of marine organisms may be abundant. Cross-bedding is frequently seen in coarse-grained sand, and ripple marks in fine-grained beds. This rock type is represented by the upper 20 to 40 feet of Mount Simon Sandstone, the Ironton Sandstone, the Sunset Point Sandstone, and the sandy base of the Glenwood Formation (plate 5).

Type 3 is characterized by fine-grained sediments and consists of shale, silty or shaly sandstone, or fine sandstone, which is usually thin-bedded. Generally it formed on a marine shelf out from the beach, seaward of Type 2, but some may have formed on tidal flats or in very shallow waters. Ripple marks and cross-bedding are common. Limy material often occurs as a cementing ingredient or as thin beds. Pale green clay and dark green pellets of the mineral glauconite are abundant in places. The fossils present usually are fragments of brachiopod shells, trilobite impressions, and burrows and trails (fig. 1.14, 1.15). According to the present interpretation, these Type 3 rocks are represented by the Eau Claire and Lone Rock formations, and the upper parts of the Sunset Point, New Richmond and Glenwood units (plate 5). They are thin and poorly developed in the Jordan and New Richmond cycles. Sometimes the reworked Type 2 rocks are

missing in the sequence, and the shaly Type 3 rocks rest directly upon the uniform quartz sand surface.

Type 4 in Wisconsin is dolomite, or shaly or sandy dolomite. (Most of the so-called limestones in Wisconsin are dolomite.) The material forming this rock was deposited in shelf-reef environments at a range of depths, from locations seaward of Type 3 shaly zones, to locations near or at the shoreline where conditions were favorable for reef-forming organisms and the collecting of deposits—tidal areas, pools, and lagoons. The limestone-forming deposits came from the reefs, the hard parts of marine animals, and chemical precipitation. If there was an influx of sediments from the mainland, a less-pure limestone resulted.

Type 4 rock is the most easily recognized of the four. Its contact with an underlying shaly unit is usually sharp and even. There are likely to be sand grains and minor amounts of shale, silt and green glauconite pellets in the lower parts of some units, but higher in the units these constituents may be completely absent. This type is represented by the St. Lawrence Dolomite, Oneota Dolomite, Willow River Dolomite, and the Sinnipee Group (plate 5).

In the rest of this chapter Wisconsin's more distinct and recognizable rock strata will be individually described. Plate 5 lists them, as well as less distinctive rock units that are not discussed. Readers should refer to it as the different rock units are mentioned. That chart has been often revised, and will be revised many times more, but currently it is the outline endorsed by the Wisconsin Geological and Natural History Survey.

A rock unit's place in a cycle will be given when known or generally agreed upon. Because of a lack of information regarding the identity and placement of some units, we are not concerned here with outlining complete cyles or giving every unit a "type" identification or position in a cycle. Geologists do not agree on the number of cycles that affected Wisconsin. The following discussion is based on a working hypothesis used by the Wisconsin Geological and Natural History Survey for continuing the study of the stratigraphy of Wisconsin's Paleozoic rocks.

The Cambrian Strata

This was the scene in Early Cambrian time. The Precambrian crystalline shield, which had been subjected to long erosion, was worn fairly smooth, with some ranges and rounded hills rising as monadnocks above the general plains. The seacoast lay far to the east and south of Wisconsin and gradually moved toward it (as though the Gulf of Mexico gradually spread northward over the southern coastal plain). The sea's encroachment was very slow. Sand was deposited along its coast as it came, at beaches and in shallow waters near the shore. A thick layering of overlapping sandstone beds can be traced from Tennessee to Wisconsin. Sand was being deposited in Tennessee in Early Cambrian time and in Wisconsin in Late Cambrian, tens of millions of years later—an estimated 530 million years ago. The formation can be traced laterally into Minnesota, Iowa, and southern Michigan.

This blanket of coalescing sand bodies that filled in and covered the old Precambrian surface is apparrently the oldest Paleozoic formation in Wisconsin. It is the *Mount Simon Sandstone,* named for a hill in Eau Claire—Mount Simon—which is its type location (fig. 4.9).[2] Sand was obviously abundant in the area at the time of deposition. Probably much of it came from the shield's disintegrating granites and quartzites.

Mount Simon Sandstone is a thick unit, up to about 1,300 feet (400 m) thick in parts of southern Wisconsin. It is Type 1 rock—moderately sorted or well-sorted quartz sand—but its sand grains are generally coarser than those of most Type 1 sandstones in the state. Thin layers of shale are found in it locally. How far north onto the now-exposed shield the Mount Simon originally extended is not known, for erosion wore back that and succeeding formations to their present irregular, semicircular position around the west, south and east sides of the Wisconsin Dome.

At its type location, where the rocks can be clearly seen in cross-section, the Mount Simon grades upward from well-sorted, thick-bedded, predominantly coarse-grained sandstone in its lower part (Type 1) to fine-grained, thinner-bedded, transitional beds of poorly sorted and well-sorted sandstone at the top (Type 2). That gradation indicates how the sea advanced and deepened over Wisconsin. As the waters over that area became somewhat quieter than along the beach zone, the sea-floor deposits included finer materials. Nearby, other good cross-sections of Mount Simon can be seen in Irvine Park on the north side of the city of Chippewa Falls, and along a service road south of Interstate Highway 94, between U.S. Highway 53 and State Highway 93.

The next formation, overlapping the Mount Simon, is the *Eau Claire,* a clayey or shaly fossiliferous sandstone of Type 3, named for its type location at the mouth of the Eau Claire River in Eau Claire. It is finer-grained and more thinly bedded than the underlying Mount Simon, and is distinguishable by the presence of abundant shale, trilobite and brachiopod fossils, and often glauconite (figs. 4.10, 4.11). The Eau Claire Sandstone thins northward from the Illinois Basin toward the Wisconsin Dome, and also from Minnesota to eastern Wisconsin, where it disappears, partly because of later erosion.

For a relatively short time in southern Wisconsin dolomite (Type 4) was forming. Here the formation is thin, but it thickens to the south into Illinois and Missouri, where the continental shelf was deeper. (There it is known as Bonneterre Dolomite.) This dolomite layer is not exposed in Wisconsin; it was found by subsurface drilling at Beloit in southern Rock County. It is overlain by a formation similar to the preceding, underlying Eau Claire, suggesting that the sea deepened when the dolomite was deposited and then became shallower again while receding. Thus the transgressive half of a cycle would have ended and the regressive half begun.

With the withdrawal of the sea from Wisconsin a period of erosion began, separating cycles of deposition. While the land stood above sea level some of the previously deposited rocks were worn away, including perhaps some of the limestone. An unconformity seen at many outcrops is taken to mark the end of one cycle of deposition and the beginning of the next.

According to the interpretation we are using, the Late Cambrian sea now slowly flooded into Wisconsin for the second time. A new succession of sediments was deposited. This new cycle began with a sandstone layer

Mount Simon
Sandstone

Precambrian

Figure 4.9 Unconformable contact of Mount Simon Sandstone with weathered Precambrian rock in a cutbank at the east side of Duncan Creek in Irvine Park, Chippewa Falls. (Wisconsin Geological and Natural History Survey)

2. Most of the units mentioned are named for locations in Wisconsin where they are well exposed or were first recognized. If type locations are outside Wisconsin or have other special significance, they will be so identified.

Information on type locations can be found in the *Lexicon of Geologic Names of the United States for 1936–1960,* Grace C. Keroher and others, U.S. Geological Survey Bulletin No. 1200 (Washington, D.C.: U.S. Government Printing Office, 1966); and in the annual *Changes in Stratigraphic Nomenclature by the U.S. Geological Survey.*

Figure 4.10 Eau Claire Sandstone's characteristically thin, alternating sandstone and weaker shaly layers are seen in these exposures south of Eau Claire. (Left photo by Jan Wiebert; right photo by M. E. Ostrom)

Eau Claire
Sandstone

Mount Simon
Sandstone

Figure 4.11 An exposure of fairly uniform beds of Mount Simon Sandstone overlain by the non-homogeneous Eau Claire Sandstone. Loose sand from the Mount Simon Sandstone is falling downslope. North of Whitehall, Trempealeau County. (Photo by M. E. Ostrom)

similar to the Mount Simon—the *Galesville Sandstone* (a member of the Wonewoc Formation; see plate 5). A bluff at Galesville in Trempealeau County, Wisconsin, is its type location (figs. 4.12, 4.13). (Some geologists believe the Galesville was left during a regressive rather than transgressive phase of a cycle.) The Galesville is thick-bedded and often cross-bedded (fig. 5.14). In many places it stands in steep slopes and forms cliffs that are striking in appearance, being usually white, sometimes gray, in color (4.14). The Galesville consists of mainly medium and fine, well-rounded quartz sand grains. Clay, silt and fossils are uncommon. It has the characteristics of origin in the beach and nearshore environment of strong waves and cur-

rents. The Galesville thins toward the Wisconsin Dome, as do many of the other Cambrian and Ordovician units.[3]

Galesville Sandstone possibly is comparable, at least in part, to some of the sandstones of the upper level of the Bayfield Group of the Lake Superior region. As mentioned earlier, it is not certain whether those northern sandstones are Upper Keweenawan (Precambrian) or Cambrian.

3. The Mount Simon, Eau Claire and Galesville used to be collectively called the "Dresbach" formation, but following accepted usage the Wisconsin Geological and Natural History Survey now reserves that term to denote a time period rather than a rock formation. The type location was in quarries near Dresbach, Minnesota.

Galesville
Sandstone

Eau Claire
Sandstone

Figure 4.12 Moving up the stratigraphic column, the Eau Claire thin-bedded, shaly sandstone formation is overlain by the relatively pure, thick-bedded Galesville Sandstone. At the Galesville's type location (at Galesville, southern Trempealeau County) the contrast between the Galesville and the upper beds of the underlying Eau Claire Sandstone can be seen. (Photo by M. E. Ostrom)

Figure 4.13 The unconformity, or irregular erosion surface, at the top of the hammer handle separates the Eau Claire and Galesville sandstone formations. (Photo by M. E. Ostrom)

Figure 4.14 A road cut showing Galesville Sandstone, west of Lake Delton in northeastern Sauk County near the Wisconsin Dells. Eroded sand lies in piles at the base. (Photo by M. E. Ostrom)

As we follow the concept of a transgressing sea, we see the waters advancing farther inland, leaving alternate beds of well-sorted coarse-grained and medium-grained sandstone, and poorly sorted, reworked beds of silt and sand. This set of layers is the *Ironton Member* of the Wonewoc Formation (fig. 4.15). Some exposures show it to be transitional with the Galesville, with no break between. It seems to have formed seaward of the beach zone in a location where high and low energy conditions alternated, where water movement was sometimes strong, sometimes weak. The Ironton's type location is Ironton, northwestern Sauk County, where iron used to be mined.

In western Wisconsin successive strata within this member can be traced for distances up to a hundred miles. The Ironton thins toward the Wisconsin Dome and thickens into Minnesota. Near the Mississippi River it is more than 40 feet (12 m) thick. It forms the surface at the Dells of the Wisconsin River. Locally it contains fragments and casts of fossils, and numerous burrows made by small animals that lived at the time it formed. Being a well-cemented unit, it caps escarpments and many outlier crags in western Wisconsin. It

is less well known in eastern Wisconsin, being deeply buried by drift there, as are other units (fig. 4.16).

Although the Ironton is now classified as the upper unit of the Wonewoc Formation, in the past it was included in the "Franconia Sandstone"[4] which formerly denoted the Ironton plus what is now known as the Tunnel City Group overlying the Ironton.

4. Now the term "Franconian," like "Dresbachian," is used to designate a time unit rather than a rock unit. The Franconia Sandstone's type location was in exposures near Franconia, Minnesota.

Figure 4.15 Alternating weaker and stronger layers in an outcrop of Ironton Sandstone near Tunnel City, Wisconsin, represent changing marine environments during the time the sediments were being deposited. (Photo by M. E. Ostrom)

The *Tunnel City* Group consists of the Lone Rock and Mazomanie formations. The *Lone Rock* is composed of shaly, fine-grained, thin-bedded and medium-bedded sandstone and greensands (sands with glauconite). Its fine particles indicate low-energy water conditions for the most part. There are abundant burrows and trails, signs of prolific animal life in this fairly quiet depositional environment. The Lone Rock is comparable to the Eau Claire Sandstone of the previous cycle.

The *Mazomanie* is a tongue of sandstone partly enclosed by the Lone Rock. It extends south from the Wisconsin Dome along the axis of the Wisconsin Arch to the vicinity of Madison. On its margin it is fine-grained, thin-bedded, non-shaly, and slightly glauconitic; it grades to medium-grained and fine-grained,

Figure 4.16 Ironton Sandstone, the surface rock at Wisconsin Dells gorge, is considerably more resistant than the sandstone underlying it, so it protrudes here as an overhanging shelf. Once the Wisconsin River had cut through the Ironton it was able to undercut it and widen its channel in the softer rock below. In time the protruding shelf cracks and falls off. Swallows have hollowed out nests in the soft sandstone of this cliff in the Upper Dells. (Wisconsin Dells Visitor & Convention Bureau)

well-sorted sandstone over the Wisconsin Arch. Locally, where firmly cemented, it forms conspicuous crags, including the natural bridge near Leland in Sauk County (figs. 4.17, 5.21, 5.22).

A characteristic differentiating the Lone Rock from the Mazomanie is the amount of glauconite found in each. The Lone Rock contains a considerable amount, whereas the Mazomanie contains little.

Overlying the Lone Rock-Mazomanie sandstones is the *St. Lawrence Dolomite,* named for St. Lawrence, Minnesota. That formation consists of the Black Earth Dolomite and Lodi Siltstone members.

The lower *Black Earth Dolomite* formed from shells of marine organisms and from stromatolite reefs built by algae (fig. 1.13). Locally this dolomite is silty, sandy and glauconitic. It is like a tongue or wedge extending north into the Lodi; it thins and pinches out to the north and increases in thickness southward into Illinois. The Black Earth is well developed west of Madison in hills bordering the valley of Black Earth Creek between Black Earth and Mazomanie (fig. 4.18).

Lodi Siltstone, the upper member of the St. Lawrence Formation, is a thin-bedded, tan to gray dolo-

Figure 4.18 Black Earth Dolomite was quarried at the base of this hill near Black Earth. Thin-bedded Lodi Siltstone overlies it. The sloping hill is Jordan Sandstone, and its wasting cap is Oneota Dolomite. (Milwaukee Public Museum. Photographer, G. O. Raasch)

mitic and shaly member with some fine sandstone locally. It is interpreted as an indication of the sea's withdrawal.

How far north the sea had transgressed during this cycle is unknown, because after it withdrew erosion removed an undeterminable amount of its deposits.

The next identifiable Paleozoic sea transgression apparently occurred near the end of the Cambrian. It deposited as its lowest, Type 1 formation the *Jordan Sandstone,* named for exposures at Jordan, Minnesota. (See plate 5.) It is composed of thick deposits of pure quartz sand, and may be seen in prominent outcrops along the sides of bluffs and cliffs in southwestern Wisconsin. Its basal member, the *Norwalk,* is fine-grained, thin-bedded to medium-bedded, well-sorted sandstone, locally containing many burrows. It is overlain by a coarser-grained sandstone, the *Van Oser Member* (thought to be named for exposures on that creek in Scott County, Minnesota), which is light brown or yellow and thick-bedded (figs. 4.19, 4.20).

Between the Van Oser Sandstone and overlying Type 4 dolomitic rocks of the Ordovician Period is a transitional, complex zone. How the rock units in that intermediate stratigraphic position should be named and classified, and where among them the Cambrian-Ordovician boundary should be placed, are unresolved and debatable problems. These transitional rocks are found in a wide area of western Wisconsin and extend into Minnesota, Iowa and Illinois. They are a heterogeneous layering of alternating and merging beds of fine-grained to medium-grained, poorly sorted dolomitic sandstones, sandy dolomites, and associated rocks of various compositions. (In early literature they have been interestingly referred to as "beds of passage.") They show the change from the predominantly sandy deposits of the Cambrian to the predominantly limy deposits of the Ordovician.

Figure 4.17 In this bluff west of Leland (Sauk County) the Mazomanie Formation overlies the Wonewoc. The light-colored, thick-bedded Galesville Sandstone member is a striking contrast to the darker overlying rock. (Photo by F. T. Thwaites. Wisconsin Geological and Natural History Survey)

Figure 4.19 Two members of Jordan Sandstone: The Norwalk—finer-grained and thinner-bedded—underlies the coarser-grained, thicker-bedded Van Oser. South of Whitehall, central Trempealeau County. The hammer provides scale. (Wisconsin Geological and Natural History Survey)

Figure 4.20 Bluffs overlooking the Mississippi River floodplain about a mile north of Victory, Vernon County. The more homogeneous lower rock of the foreground exposure is Jordan Sandstone. The overlying rock with thinner varying beds is a transitional member, leading up into the Prairie du Chien Dolomite. This is approximately the Cambrian-Ordovician boundary. (Milwaukee Public Museum)

The name that tentatively was given to these transitional rocks overlying the Van Oser was *Sunset Point*. However, recent research indicates that the name should be used only locally for Sunset Point Sandstone in the vicinity of its type location, which is the hill known as Sunset Point in Hoyt Park, Madison; there the rock is exposed in former quarries. Sunset Point Sandstone appears not to overlie the Van Oser, but to be its lateral equivalent. It is a fine-grained, cream-colored sandstone which in past years was called "Madison Sandstone" and was used in the construction of many early buildings of southcentral Wisconsin, including North, South and Bascom halls and other buildings on the University of Wisconsin–Madison campus.

The proposed substitute name for the transitional rocks throughout western Wisconsin is *Coon Valley*, after that town in western Vernon County.

The location of the Cambrian-Ordovician boundary within this transitional zone remains indefinite because in the Wisconsin area there seems to have been no interruption in sedimentation or in the fossil sequence at the close of the Cambrian. There were just gradual, fluctuating changes in the kinds of sediment deposited. Geologists are trying to correlate fossils in these rocks with fossils in rocks elsewhere that have already been designated as Cambrian or Ordovician, in order to relate the Wisconsin area chronologically and stratigraphically to other areas.

An Overview of Wisconsin's Cambrian Sandstones

Upper Cambrian formations in Wisconsin, which have been described, are often referred to collectively as "the Cambrian sandstones." They form the bedrock surface of more than a quarter of the state. In the southern counties they are thickest, but to the north they thin and fray until they disappear on the northern Precambrian dome (plate 1).

Where they are not covered by Ordovician rock in Wisconsin these sandstones are exposed in a band that has been described as a "crescent" around the western, southern and eastern sides of the dome. This crescent is the *Cambrian sandstone lowland,* sometimes called the "Central Plain" (plate 4). Its eastern and northwestern parts are covered by glacial drift. Many picturesque sandstone mounds dot its unglaciated section.

The eastern "horn" of the shaggy-edged crescent, the narrowest part, is highly irregular in width, with a maximum width of about 25 miles (40 km). Toward the north, where the horn is most segmented by erosion, it has left large outliers. (An outlier [first mentioned in chapter 1] is a portion of a stratified formation that stands detached, away from the main body; the portion that formerly connected them has been removed by erosion [fig. 5.3].) The crescent's western "horn," which is wider than the eastern one, extends north past Hayward to the Bayfield County boundary. The middle part of the crescent—the widest part—takes in essentially all of Waushara, Marquette, Adams, Juneau, Jackson, and Eau Claire counties, and a considerable area in counties adjacent to them. This main segment of the Cambrian sandstone lowland ranges in width from 40 to 70 miles. The plain is described in more detail in the following chapters.

Cambrian sandstones are exposed in places outside the crescent where erosion has removed overlying rocks, as in the Madison "Four Lakes" area, along the Lower Wisconsin River, and in valley floors of rivers north of the Lower Wisconsin. The La Crosse and Black rivers have been especially effective in removing overlying rock and exposing the Cambrian sandstones near the Mississippi River. The Cambrian sandstones are seen also on the "nose" and "mouth" areas of the Indian Head in the northwest; and because of their easily studied exposures in the St. Croix River valley, they have been named the *St. Croixan Series*. This series, a time unit, includes all of the Upper Cambrian sedimentary formations in the Wisconsin area, from the Mount Simon up through the Jordan (plate 5).

Before there had been much study of Wisconsin's Cambrian sandstones, they were collectively referred to as "Potsdam Sandstone," since they were considered equivalents of the Potsdam rocks of New York State (fig. 4.21).

In generalizing, it can be said the St. Croixan (Upper Cambrian) Series of rocks consists mainly of sandstones with interbedded shales, siltstones and dolomites. The sandstones in central and southern Wisconsin are commonly light in color—buff, yellowish or white, and varying locally to shades of pink, brown and green. (Sandstones of the Lake Superior region, some of which may be Cambrian, are either light-colored or reddish brown.) The green mineral glauconite occurs in beds scattered through the series, and where it is concentrated it forms the sandstone called "greensand." Greensands can be seen south of Lone Rock and elsewhere along the Wisconsin River in Iowa County; at road cuts west of Cross Plains in Dane County, along

U.S. Highway 14; at road cuts in Vernon county on U.S. Highway 14, and on State Highway 35 along the Mississippi River; and at many exposures in Juneau and Monroe counties.

Wisconsin's Type 1 Cambrian sandstones are known for their unusual purity and textual maturity. Their quartz content is high; they are relatively free of silt and clay. The sand grains are exceptionally well rounded. These sands are thought to have been repeatedly recycled through deposition, erosion, and redeposition in Early and Middle Cambrian time, before having been reworked and redeposited during Late Cambrian.

The Ordovician Strata

There was no hiatus in the deposition of sediments as the Ordovician Period began. (Refer to plate 5.) But there was a transition from one kind of rock to another—from the homogeneous, pure quartz Jordan Sandstone to the dolomitic *Prairie du Chien Group* (fig. 4.22). Here is introduced the first major Paleozoic se-

Figure 4.22 Table Rock, northwest of Readstown in southern Vernon County, has also been called Five-Point Rock. Its "legs" are Jordan Sandstone, and its "tabletop" is the more resistant, dolomitic Prairie du Chien rock. (Wisconsin Department of Natural Resources. Photographer, F. G. Wilson)

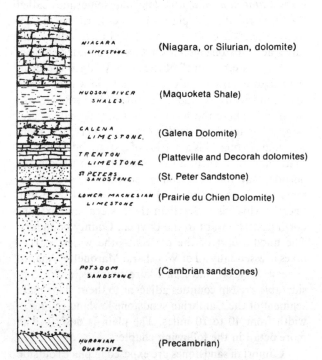

NIAGARA LIMESTONE	(Niagara, or Silurian, dolomite)
HUDSON RIVER SHALES	(Maquoketa Shale)
GALENA LIMESTONE	(Galena Dolomite)
TRENTON LIMESTONE	(Platteville and Decorah dolomites)
ST PETERS SANDSTONE	(St. Peter Sandstone)
LOWER MAGNESIAN LIMESTONE	(Prairie du Chien Dolomite)
POTSDAM SANDSTONE	(Cambrian sandstones)
HURONIAN QUARTZITE	(Precambrian)

Figure 4.21 A stratigraphic column of Paleozoic rocks of southern Wisconsin published in 1900 with names of formations used at that time. Current equivalents are in parentheses. Compare to Plate 5. (Wisconsin Geological and Natural History Survey, Bulletin No. 5)

quence of carbonate, or dolomite, rocks in Wisconsin. They formed in shallow waters which were not receiving sand and silt—in quiet waters some distance seaward from shore, in reef areas, and at barely wet shorelines. The relatively small amount of sand and finer particles in the rock indicates that the above-water parts of nearby landmasses had decreased or that the land had submerged.

The Prairie du Chien Dolomite, which originally must have overlain more of the Cambrian rocks than it does now, has been eroded back to where it forms the outer (southern) edge of the Cambrian sandstone crescent (plate 1). In northeastern Wisconsin its exposed area is just a narrow band crossing the Menominee River near McAllister (Marinette County). The Prairie du Chien Dolomite extends southwest into the state parallel to, and east of, the east horn of the sandstone crescent, through the towns of Coleman, Oconto Falls and Medina, and Rush Lake in southwestern Winnebago County. It widens in the southcentral part of the state, and follows around the south and southwest sides of the crescent in broken fashion. Its largest single broad area is the "jaw" of the Indian Head in St. Croix and Pierce counties (fig. 4.23).

The *Oneota Dolomite Formation* (named for Iowa's Oneota River, now called the Upper Iowa River) is the lower and main body of the dolomite in the Prairie du Chien Group. It includes all of the dolomitic quartz sandstone, sandy dolomite and dolomite above the Jordan Sandstone and below the New Richmond Sandstone of the overlying *Shakopee Formation* (named for outcrops at Shakopee, Minnesota).

The Oneota is easy to recognize, being a massive, buff-colored body of rock with quite different strata above and below. It contains abundant algal structures and oolites. *Oolites,* which resemble fish eggs in form, are sand grains or other tiny particles coated with a

Figure 4.23 Prairie du Chien Dolomite is the caprock on many valley bluffs in the Lower Wisconsin River Valley, as here near Avoca, Iowa County. (Gene Musolf)

hard substance, silica, that was chemically precipitated from sea water. The Oneota is the first Paleozoic unit thus far to contain chert. Although the Oneota's lower part is sandy, its upper part is mainly pure dolomite.

There is a question as to whether deposition was interrupted between the forming of the Oneota and the overlying Shakopee strata. At most exposures there are signs of erosion between the two formations. On that basis—if deposition was not continuous—the lower member of the Shakopee, the *New Richmond Sandstone,* would represent the start of another cycle of deposition, and would be considered the coalesced bodies of beach and nearshore sands laid down by an encroaching sea. The New Richmond is in part quartz sandstone, but it blends vertically, and laterally to the southeast, to include interbedded sandy dolomite and gray-green shale. It is only about 5 feet (1.5 m) thick in southcentral Wisconsin, and thickens westward to about 20 feet (6 m) in southwestern Wisconsin (fig. 4.24).

Shakopee Formation

Willow River Dolomite

New Richmond Sandstone

Oneota Formation

Figure 4.24 New Richmond Sandstone is not always as uniform appearing as in this quarry east of Eastman in Crawford County. Overlying it is the Willow River Dolomite. Underlying it is the Oneota Formation which was eroded prior to deposition of the sandstone. (Photo by M. E. Ostrom)

The upper member of the Shakopee Formation (and uppermost of the Prairie du Chien Group) is the *Willow River Dolomite,* which overlies the New Richmond Sandstone along a sharp, even contact. It is named for exposures at Willow River, St. Croix County. This dolomite formed when water had become deeper throughout the entire Upper Mississippi Valley area. It is a sequence of interbedded sandy and oolitic dolomites with local thin beds of chert, gray-green shale, and quartz sandstone; and it is similar to the Oneota below the intervening New Richmond Sandstone.[5]

In general, the Prairie du Chien dolomites are light in color—light buff, gray, off-white, and sometimes white. They commonly contain flint or chert nodules. Some of the rock is brecciated; it was broken up by waves, the pieces were rolled about, and afterward they were recemented.

The Prairie du Chien dolomites contain many cavities, which were probably caused by percolating groundwater dissolving the lime. The smaller cavities are usually completely filled with crystalline quartz that was precipitated from groundwater, but larger ones are hollow in the center and crystal-lined (fig. 4.25). Geologist T. C. Chamberlin described them in 1877 in his pictorial prose:

> Where the cavities are larger, the crystals only form a lining, producing drusy little grottoes, some of which are very beautiful. The quartz is most frequently transparent or opalescent, but it is sometimes red, brown, or rose colored. The crystals are sometimes grounded on a chalcedonic base, forming a beautiful combination.[6]

Cavities of cave size formed in the dolomite too, as the moving groundwater slowly dissolved and removed the rock.

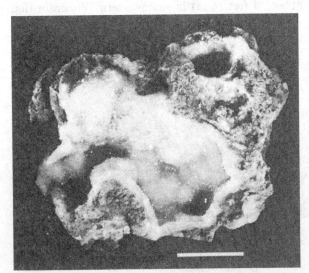

Figure 4.25 Quartz crystals lining a cavity in Ordovician dolomite, from Grant County. Line scale = 2 cm (0.8 in.). (Geology Museum, University of Wisconsin-Madison. Photographer, Lawrence D. Lynch)

The thickness of the Prairie du Chien Group is uneven, determined to a considerable degree by the amount of erosion that followed emergence from the sea. That erosion period was a pronounced one. The sea made a full withdrawal to the south and east of Illinois, exposing more land in the Upper Mississippi Valley area than at any time since the first Cambrian sea had arrived. The surface of the Prairie du Chien became highly dissected by erosion and many deep valleys were carved into it. Some were so deep that they cut down through the Oneota and most of the Cambrian rock to the lowest formation, the Mount Simon. The undulating irregularities on this surface reminded T. C. Chamberlin of rising and falling ocean swells, and he referred to them metaphorically as "petrous billows."

Another cycle of deposition began as the Ordovician sea returned to Wisconsin. Over that "wavy" unconformity the sediments that became the *St. Peter Sandstone* were laid. Where they filled valleys they were thick accumulations, and where they just covered the "crests," the higher land between the valleys, they were thinner. As a result, there is great variation in the thickness of St. Peter Sandstone, even within short distances. In a few places it measures more than 300 feet (90 m) thick. In some places it is completely absent. At the forefront of the encroaching sea, the St. Peter sediments seem to have been deposited in a variety of nearshore locations, including the beach and underwater shelf, and perhaps in lagoons. Commonly they settled in the form of complex undulations, ridges and dunes, some more than 30 feet high. The sand grains appear to have been sorted and polished by wind; at some time they may have been a desert sand.

St. Peter Sandstone was named for a river in Minnesota, now called the Minnesota River. It is the main sandstone stratum in the predominantly dolomitic Lower Ordovician sequence of rocks.

In most places St. Peter Sandstone is so pure and so free of cementing material that is crumbles easily, and often is, as Chamberlin wrote, "little more than an incoherent sand bed." But when exposed to weathering its surface becomes hard, especially where impregnated with iron (as the "red rock" below Darlington in Lafayette County). Some weathered St. Peter is strong enough to stand in high bluffs or mounds. In Governor Dodge State Park, central Iowa County, it forms most of the hills, ridges and sheer cliffs. The hardened rock is sometimes used as building stone, though it is not particularly durable (fig. 4.26).

5. The whole Prairie du Chien Group used to be called "Lower Magnesian limestone"—"magnesian" because of its magnesium content, and "lower" to distinguish it from the "Upper Magnesian limestone" now called Sinnipee and Niagara dolomite (fig. 4.21).

6. T. C. Chamberlin, *Geology of Wisconsin. Survey of 1873–1877* (Madison: Commissioners of Public Printing, 1877), vol. 2, p. 269.

Figure 4.26 An eroding crag of St. Peter Sandstone near Pine Bluff, western Dane County. Pieces of fallen rock and sand from the disintegrating sandstone start their trip downslope. (Gwen Schultz)

Figure 4.27 Quarry in St. Peter Sandstone at Klevenville, western Dane County. This is the Tonti member—thick-bedded and uniform in texture. The pure, well-sorted sand that such rock provides is much in demand. (Photo by M. E. Ostrom)

St. Peter Sandstone is typically white, gray or light buff. Being light in color, it is visible from long distances. It is conspicuous along the Mississippi River in Grant County. Where iron-stained, it is reddish or yellowish. Some of it has a pink or green cast, and in places groundwater solutions carrying coloring minerals have given it striking irregular bands and variegated spotted markings. It is generally lacking in fossils.

In eastern Wisconsin the St. Peter, largely drift-covered, is a narrow strip east of the stronger Prairie du Chien dolomites (plate 1). From Marinette County it runs southwest, west of Lake Winnebago, past Lake Poygan and on into Jefferson and Dane counties, where its distribution is broader and irregular. Throughout southwestern Wisconsin streams have cut through overlying rocks and exposed the St. Peter along their branching courses. Because it erodes easily it is commonly found under a ledge of more resistant rock, and on geologic maps may appear as just a narrow ribbon along tributaries; but it remains on the valley floors of some rivers.

The best exposures of St. Peter Sandstone are in major river valleys from the Rock River west, particularly in the Sugar River valley. North of the Wisconsin River the St. Peter has been largely eroded away, but it does remain in sizable areas on the higher divides near the Mississippi River north of Prairie du Chien (through Crawford County to northcentral Vernon County), and overlying the "chin" part of the Prairie du Chien Dolomite "jaw" of the Indian Head.

In some places the St. Peter has a shaly to conglomeratic zone at its base—the *Readstown Member.* Overlying that base is the widely distributed main member, the *Tonti,* named for exposures in Tonti Canyon, Starved Rock State Park, Illinois. The Tonti is a thick-bedded body of well-sorted, pure, fine-grained and medium-grained, light-colored quartz sandstone, extremely uniform in texture (fig. 4.27).

The *Glenwood Formation* (named for exposures in Glenwood Township, Iowa) overlies the main Tonti Member of the St. Peter Sandstone. The thin formation consists of sandstone, shaly sandstone, and dolomitic and silty shale. In Wisconsin it ranges in thickness from less than a foot in the vicinity of the Wisconsin Arch in southern Dane County to about 14 feet (4 m) in western Grant County. It is transitional between the well-sorted, uniform St. Peter Sandstone below and the dolomite above, and represents a brief episode of significant change. The incoming sea reworked the upper surface of the St. Peter sands. As new deposits were laid down on the sand, they too were disturbed and redeposited. The Glenwood, grading upward into shale and then into the overlying dolomite, illustrates the effect of deepening waters as the sea's shoreline shifted far to the north (figs. 4.28, 4.29).

In Wisconsin, the next group of dolomites begins with the *Platteville Formation;* it is followed by the *Decorah* and *Galena* formations, in ascending order, which are named for cities in Iowa and Illinois, respectively.[7] In current terminology the Platteville, Decorah and Galena are combined as the *Sinnipee Group* in Wisconsin, which is named after a bluff north of Sinnipee Creek, Grant County, where all three formations are exposed. These three carbonate formations are related in that together they are all distinctly different from the underlying and overlying rocks. However, they are distinguishable from one another on the basis of texture, impurities, bedding, fossils, and, in some places, color.

7. These formation used to be referred to collectively by compound names like "Galena—Black River Dolomite" or the "Plattteville—Galena Group"; and the Platteville and Decorah used to be called the "Trenton" limestone formation (fig. 4.21). The term "Platteville Limestone," also found in older classifications, covered a wide stratigraphic latitude; it included all the beds between the top of the St. Peter Sandstone and the base of the Galena Dolomite.

Soil and
weathered rock

Platteville
Dolomite

Glenwood
Formation

St. Peter
Sandstone

Figure 4.28 A cross-section showing the gradation from St. Peter Sandstone to Platteville Dolomite, south of Beetown in western Grant County. (Photo by M. E. Ostrom)

Figure 4.29 Platteville Dolomite overlying the Glenwood Formation in a hillside along U.S. Highway 151 in northwestern Lafayette County. The weaker Glenwood, partly sandy and shaly, is transitional between underlying St. Peter Sandstone and the dolomite shown. (Wisconsin Geological and Natural History Survey)

On the bedrock geologic map the Sinnipee Group of dolomites appears in eastern Wisconsin as an irregular band running from the Michigan border along the west side of Green Bay, and south through the Lake Winnebago-Rock River lowland into Illinois (plate 1). Drift covers the rock there, of course. In driftless southwestern Wisconsin these dolomites stand out on the map as the main surface rocks south of the Wisconsin River. There are outliers north of the river, as well as in the glaciated Indian Head "jaw," where they overlie the St. Peter Sandstone in western Pierce and St. Croix counties (fig. 4.30).

In eastern Wisconsin the Sinnipee rocks are the floor of the north-south lowland, but in southwestern Wisconsin they form highlands. There the maximum thickness of the Platteville Dolomite is about 75 feet

Figure 4.30 A geologist of an earlier generation studies an outcrop of the lower Platteville Dolomite in western Grant County northeast of Cassville. (W. O. Hotchkiss. Wisconsin Geological and Natural History Survey)

(23 m). The Decorah is about a third as thick, thinning eastward onto the Wisconsin Arch. It contains an abundance of fossils, and is composed of gray shelly dolomite and thin beds of green shale. It grades upward into the Galena, the thickest formation of the Sinnipee Group, which has a fairly uniform thickness of about 225 feet (70 m) in southwestern Wisconsin.

Wisconsin's main zinc and lead deposits have been found in the Galena and Decorah formations and in the upper part of the Platteville in Grant, Iowa and Lafayette counties. There is about ten times more zinc than lead in the deposits. The process by which ores of these metals formed in veins and pockets in the rock is not fully understood (fig. 4.31).

Galena Dolomite is the chief formation containing lead ore, or galena, in this area. It is a thick-bedded and unusually homogeneous stratum. In southwestern Wisconsin it is gray or buff-colored; in eastern Wis-

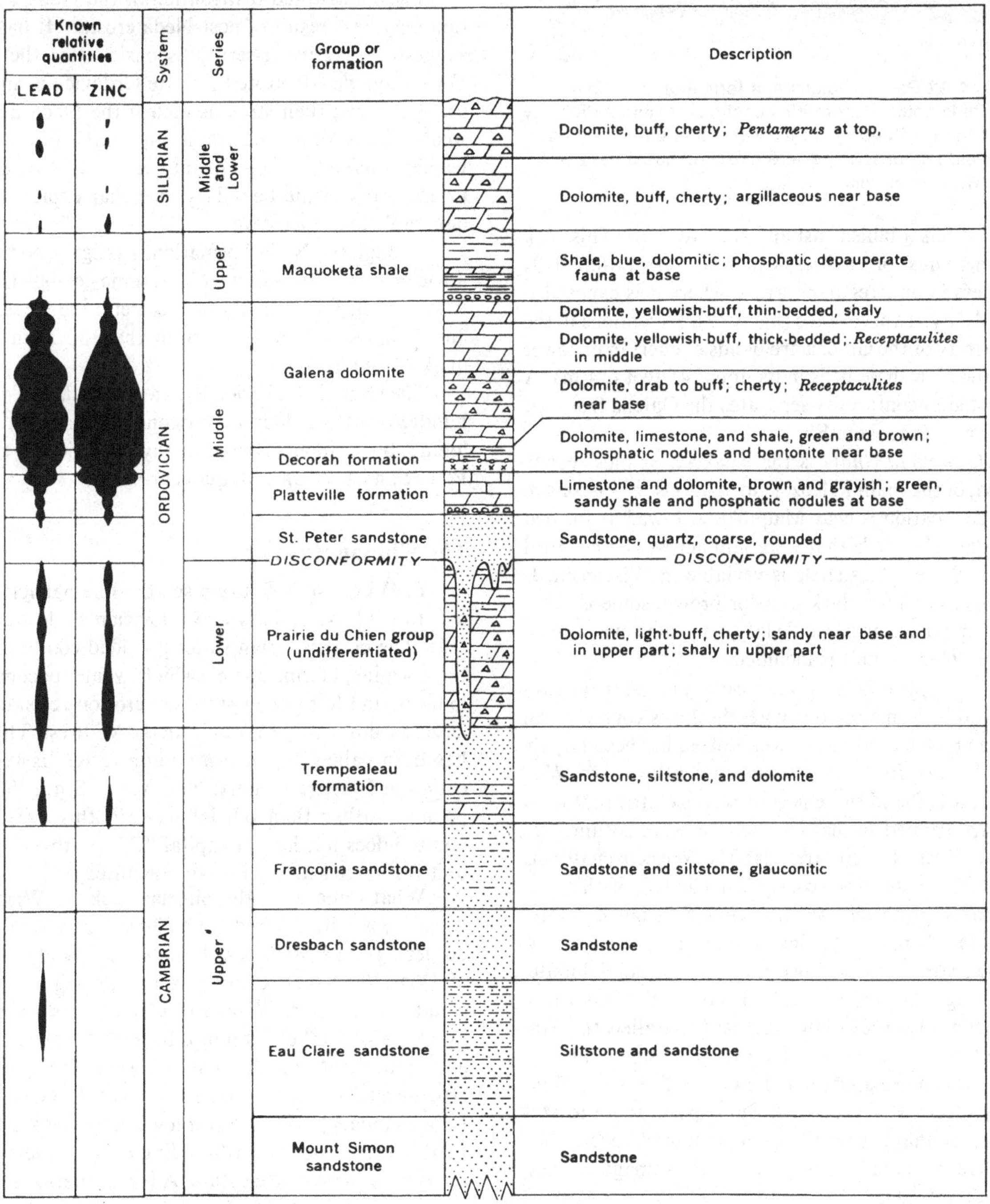

Figure 4.31 Simplified stratigraphic section showing relative quantitative stratigraphic distribution of lead and zinc in the Upper Mississippi Valley district. (From Heyl and others. U.S. Geological Survey, and Wisconsin Geological and Natural History Survey)

Figure 4.32 Galena Dolomite, a formation containing lead ore in southwestern Wisconsin, as seen on Highway 151 at the south edge of Platteville. It is thick-bedded and homogeneous, and where exposed has a decayed look. (W. A. Broughton)

consin it has a bluish cast and is more shaly. This rock contains vugs, or cavities, which are crystal-lined. It weathers in an irregular way, so where it is exposed it has a decayed appearance (fig. 4.32). Exceptions to the uniformity of the Galena are bands of chert in its lower part and thin shale beds in its upper 45 feet (14 m). A suspected unconformity separates the Galena from the overlying Maquoketa Shale.

Maquoketa Shale is the weakest, or most easily eroded, of the main Paleozoic formations in Wisconsin. Its type location is near Maquoketa, Iowa.[8] It formed from vast clay beds that were laid across northcentral United States. This shale is variable in Wisconsin: It is bluish, greenish, dark gray, or brown; some is quite soft, some more brittle and slaty, and some even sandy. A thin dolomite unit is included.

This Upper Ordovician formation weathers rapidly, and so forms relatively gentle slopes covered with wasting rock material. The Maquoketa has been largely eroded away from the surface in southwestern Wisconsin; it is found there now only as isolated remnants that are capped by harder rock, or were so until recently. Most of these are near the Wisconsin-Illinois border. In eastern Wisconsin what is left of this formation is protected by overlying Silurian dolomite (plate 1). There the Maquoketa shows as a long, narrow, winding strip below and west of the Silurian dolomite, extending from the east side of Green Bay (at Little Sturgeon Bay) south into Illinois. It parallels the Sinnipee Group, which lies along its west side, and thus makes up the east edge of the Green Bay-Lake Winnebago-Rock River lowland. But exposures of the Maquoketa in that glaciated region are few and poor. They are usually found in fresh road-cuts, drainage ditches, and other diggings.

Overlying the Maquoketa Shale, but only in places, is the thin and interrupted iron-bearing *Neda Forma-*

tion. It is found in separated sites in eastern Wisconsin from Door County south along the eroding western edge of the overlying Silurian rocks. Its iron-rich beds occur as horizontal lens-shaped deposits; they span areas measured in square miles, and are from a few feet to more than 50 feet thick. The deposits consist of hematite with red shale layers and shale pebbles. The hematite, which is about 45 percent iron, is in the form of tiny, flattened grains (oolites); these have been described as "shot ore" or "flax seed ore."

The interrupted distribution of the Neda Formation may be a result of post-Neda erosion. It has been suggested that in eastern Wisconsin after the Neda Formation was deposited in Late Ordovician time the sea withdrew; then streams etched the Neda and the weak Maquoketa beneath it into hills and valleys, leaving an uneven unconformity at the Ordovician-Silurian stratigraphic boundary. But that explanation is debatable and uncertain.

Though the Neda Formation is assigned to the Ordovician on the basis of its fossils, some geologists think it was formed as the Silurian sea returned, reworking and redepositing material from the underlying Maquoketa Shale.

The Ordovician Period left an accumulation of rock hundreds of feet thick in Wisconsin. Most of it is dolomite, but it includes noteworthy exceptions—the St. Peter Sandstone and Maquoketa Shale.

The Silurian Strata

In Wisconsin's Silurian sea the fine particles that had formed the underlying Ordovician shale were absent. The clear Silurian water provided conditions for the forming of limestone (which would become dolomite), and left one of the most erosion-resistant Paleozoic sedimentary rock units in the Midwest. This unit has been called *Niagara dolomite* (after its type location in Niagara County, New York). Since that is a popular rather than official term in this area, "dolomite" does not have a capital "D." At this time the unit is referred to as Silurian dolomite.

What remains of the Silurian rocks in Wisconsin is mainly a drift-covered band 230 miles (370 m) long, bordering Lake Michigan. From Rock Island at the tip of Door Peninsula it curves south through Kenosha county and eastern Walworth County, and continues into Illinois. To the east it dips beneath Lake Michigan and into the Michigan Basin beyond. The western eroding edge of the Silurian rocks faces the Green Bay-Lake Winnebago-Rock River lowland of weaker rocks. In Illinois, the southward-eroding edge of these rocks turns west through that state. A few outliers remain in

8. In older reports Maquoketa Shale was called "Richmond" or "Hudson River" or "Cincinnati" shale.

Figure 4.33 An abandoned quarry of Niagara dolomite at High Cliff east of Lake Winnebago. (Wisconsin Department of Natural Resources)

driftless southwestern Wisconsin on top of some of the higher hills, including Blue Mound, and the largest Platte mounds which are several miles northeast of Platteville (plate 1, figs. 4.33, 5.48).

In trying to picture the former extent of the Silurian sea and its rocks in Wisconsin, one considers that they once lay over what is now the highest point in southern Wisconsin—Blue Mound in eastern Iowa County, whose elevation is 1,719 feet (524 m). Farther north, chert pebbles that may be Silurian have been found high on the Baraboo Hills. Envisioning the Silurian rocks as they originally were—continuous from Illinois over Blue Mound and possibly across the Baraboo Hills—one can surmise that they probably extended some distance beyond.

Plate 5 shows units within the Silurian System and their current classification, which is being revised. Taken together, the units have a maximum thickness of about 600 feet (180 m).

The *Mayville* is the lowest and best exposed Silurian formation in Wisconsin. It is predominately a coarse-textured, cherty, locally brecciated body of buff or gray dolomite, with a maximum thickness of about 175 feet (53 m). From Door Peninsula to Illinois it is present in the irregular, north-south line of craggy cliffs and escarpments just east of the Green Bay-Lake Winnebago-Rock River lowland. It is found also in outliers of southwestern Wisconsin.

The Silurian dolomites formed mainly from reefs and material deposited around them, as described in chapter 1. In quarries and other exposures remains of

reefs and associated marine organisms can be seen: some coral masses standing erect, as they grew; others broken into fragments; and fossilized remains of trilobites, crinoids, mollusks, and other forms of sea life. Some of the calcareous material was ground by waves to a fine powder and spread evenly over the sea floor, and through time this white mud was converted to a beautiful, compact, pale, fine-textured dolomite of the kind seen in the Byron beds.

The *Byron Dolomite* (named for Byron Township in Fond du Lac County) presents a marked contrast to the coarse-textured Mayville beneath. The Byron forms bold, picturesque cliffs on the east side of Green Bay from Little Sturgeon Bay north to the end of Door Peninsula. Pairs of such cliffs stand at the mouth of Big Sturgeon Bay and at harbors to the north. This fine-textured rock also forms white cliffs south of Fond du Lac. Some Byron Dolomite has a gray or cream-colored tint and is lined or mottled with pink, making it a handsome ornamental stone; and some near Brillion in northeastern Calumet County is hard enough to be polished to a fair, marblelike luster.

The *Racine Dolomite* deserves special mention because of its extensive reef devlopment and its rich diversity of fossils. It is generally a thick-bedded, pure dolomite, and includes the high-quality "Lannon stone" of Waukesha County (fig. 4.34). The Racine area of Wisconsin is one of the Midwest's most famous Silurian locations. There fossiliferous Racine Dolomite is exposed in quarries and along Lake Michigan at Wind Point. Beneath the lake water a mile southeast of the Racine harbor entrance is a bedrock structure that is named "Racine reef" on nautical charts. It is probably an old Silurian reef that was exhumed by erosion and then resubmerged in lake water.

Figure 4.34 A "Lannon Stone" quarry at Sussex, northeastern Waukesha County. (Ralph V. Boyer)

The Devonian Strata

A period of erosion separated the Silurian and Devonian periods. It was not until Middle Devonian time that the sea returned to Wisconsin.

There is not much Devonian rock in Wisconsin—just a remnant strip along the Lake Michigan shore from Milwaukee to Sheboygan, which extends eastward under the lake (plate 5). There are few natural exposures of Devonian rock here because its area is small and it is drift-covered. The lack of identifiable outliers to the west leaves us without a clue to its original extent, but it surely covered much of the state. This rock was first recognized as belonging to the Devonian Period (Age of Fishes) in 1860 when samples taken near Milwaukee were found to contain fossil fish remains. During the excavation of water-intake tunnels for the city of Milwaukee from 1885 to 1900 and from 1912 to 1917 much was learned about the subsurface Devonian rocks.

Wisconsin's Devonian rocks are mainly dolomite, especially the lower formations. Some of the rocks have thin shale partings, and the uppermost formation is shale (fig. 4.35).

The oldest of Wisconsin's Devonian formations, the *Lake Church,* has the most northerly exposures, near the village of that name in northeastern Ozaukee County about 30 miles (48 km) north of Milwaukee. There it has been observed in a now-flooded quarry and, at times of low water, along the lake shore at Harrington Beach State Park.

The overlying *Thiensville Formation* has as its type location a road cut on State Highway 57 two miles north of that city in southern Ozaukee County. Both this and the Lake Church Formation are Middle Devonian gray-to-brown dolomite and are somewhat bituminous. The Lake Church contains fossils, but they are rare in the Thiensville.

The still-younger *Milwaukee Formation* appears along the Milwaukee River in Lincoln and Estabrook parks on Milwaukee's north side. Its uppermost member occurs only in the subsurface, and is known from well-drilling and tunneling. Rocks of the Milwaukee Formation are mainly grayish shale with some dolomite and limestone. From 1876 to the early 1900s they were quarried along the Milwaukee River for the making of natural cement and were referred to as "cement rock."

The youngest Devonian (and youngest Paleozoic) formation in Wisconsin is the *Kenwood Shale.* It was named for the Kenwood section of northeastern Milwaukee where it is found, but "Kenwood" had already been used for different units elsewhere, so another name may be assigned to this formation in Wisconsin. The Kenwood Shale correlates with the Antrim Shale of Michigan.

For a time the Kenwood was thought to be Mississippian in age, but it is now considered Late Devonian. It is mainly a black fissile shale. Thin, fragile chips of it have been found in glacial drift, but no outcrop has been seen; it is known from drilling and digging. Tunnel excavations near Milwaukee's lakeshore at the end of North and Linnwood avenues encountered 15 feet (5 m) of black shale above the gray shales and limestones of the Milwaukee Formation. The Kenwood's maximum known thickness is 55 feet (17 m). Several thin seams of coal have been found among these shales.

After the Devonian

Geologic time from Late Devonian to the present is not represented by any bedrock in Wisconsin. The sea left. It undoubtedly returned—probably in Late Paleozoic, perhaps even after that—but tangible proof is absent.

However, certain old gravel deposits found on and near hilltops in the Driftless Area may have import. They are known as the *Windrow gravels* because some are found on Windrow Bluff in Monroe County. Some are found also at Seneca in Crawford County, on the Baraboo Hills, and on a number of other uplands in southeastern Wisconsin, as well as in nearby sites in Minnesota, Iowa and Illinois. These scattered deposits show similarities. The gravels are largely quartz or chert—virtually insoluble; and many of the deposits are cemented with iron or other binding material, so they were long preserved. Fossils in the chert and underlying rock indicate that the gravels may be of Cretaceous age, much younger than the Paleozoic rock (fig. 1.9). Farther west similar material does occur in Cretaceous conglomerate. Some geologists think the gravel may be even younger than that.

The Windrow gravels are generally well rounded and polished, and appear to be river gravels left on valley bottoms. If this is true, then what is now hilltop was valley floor when they were deposited; and hills from which the old rivers flowed, which would have been above the level of the present uplands, were removed by erosion. Thus it could be that ocean waters covered at least the southwestern Wisconsin area in post-Paleozoic time, leaving layers of sedimentary rock, and that the old streams that eroded those rocks left evidence of the rocks' existence in these separated Windrow gravel deposits, which in turn will be washed away eventually. But this explanation of the gravels is open to question.

After the last retreat of the sea, however many millions of years ago that was, erosion continued carving into Wisconsin and wearing away its rocks. What that "sculptor" did to the Paleozoic "stone" it had to work with, what part it played in fashioning our geologic foundations, is of special significance to us today.

COLUMNAR SECTION OF THE DEVONIAN IN WISCONSIN

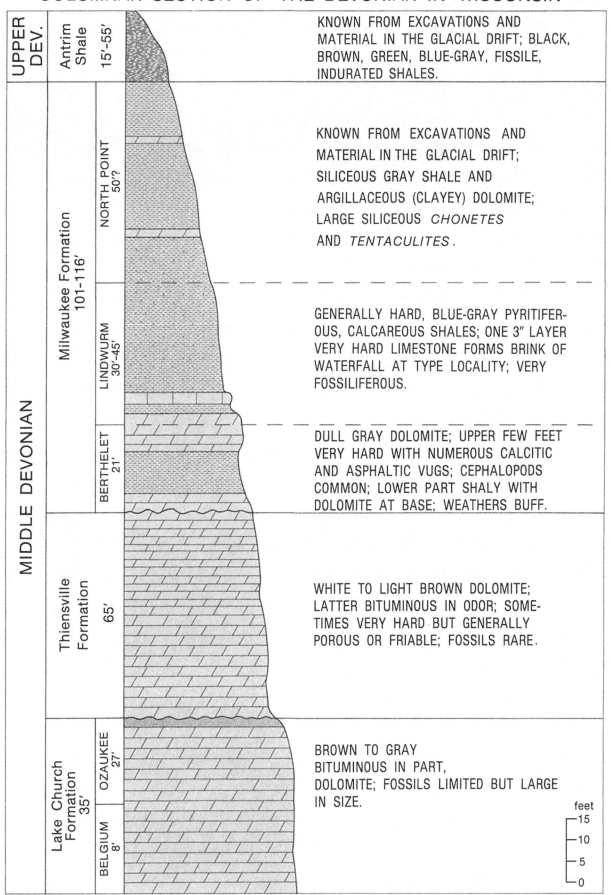

UPPER DEV.	Antrim Shale	15'–55'	KNOWN FROM EXCAVATIONS AND MATERIAL IN THE GLACIAL DRIFT; BLACK, BROWN, GREEN, BLUE-GRAY, FISSILE, INDURATED SHALES.
MIDDLE DEVONIAN	Milwaukee Formation 101–116' / NORTH POINT 50'?		KNOWN FROM EXCAVATIONS AND MATERIAL IN THE GLACIAL DRIFT; SILICEOUS GRAY SHALE AND ARGILLACEOUS (CLAYEY) DOLOMITE; LARGE SILICEOUS *CHONETES* AND *TENTACULITES*.
	Milwaukee Formation / LINDWURM 30'–45'		GENERALLY HARD, BLUE-GRAY PYRITIFEROUS, CALCAREOUS SHALES; ONE 3" LAYER VERY HARD LIMESTONE FORMS BRINK OF WATERFALL AT TYPE LOCALITY; VERY FOSSILIFEROUS.
	BERTHELET 21'		DULL GRAY DOLOMITE; UPPER FEW FEET VERY HARD WITH NUMEROUS CALCITIC AND ASPHALTIC VUGS; CEPHALOPODS COMMON; LOWER PART SHALY WITH DOLOMITE AT BASE; WEATHERS BUFF.
	Thiensville Formation	65'	WHITE TO LIGHT BROWN DOLOMITE; LATTER BITUMINOUS IN ODOR; SOMETIMES VERY HARD BUT GENERALLY POROUS OR FRIABLE; FOSSILS RARE.
	Lake Church Formation 35' / OZAUKEE 27'		BROWN TO GRAY BITUMINOUS IN PART, DOLOMITE; FOSSILS LIMITED BUT LARGE IN SIZE.
	Lake Church Formation / BELGIUM 8'		

feet
- 15
- 10
- 5
- 0

Figure 4.35 Columnar section of the Devonian of Wisconsin. (Antrim is now being called Kenwood.) (Jeannette E. Roberts)

5

The Eroding Paleozoic Rocks—
Their Meaning to Us

So far, the focus has been on Wisconsin's Paleozoic rocks from the standpoint of formative processes and structural arrangement. Now they will be considered from a different perspective—from the surface, from the observer's point of view—to see what effects these diverse rocks have had on topography, other aspects of the geographic setting, and human activities.

A few words of review and retrospect. The sedimentary rocks of Paleozoic and younger age that were layered on and around the Precambrian shield to form much of the North American continent were, of course, originally horizontal or nearly so. Later some parts of the continent, being subjected to crustal movement with pressure from below or the side, were warped slightly, or bent severely into folds, or faulted. In places the pressure was great enough to force rocks to mountain height, as in the Rockies and Appalachians. Wisconsin's Paleozoic rocks, lying upon stable Precambrian basement rocks, experienced only mild disturbance; but even just the gentle tilting of strata that took place here was enough to create significant irregularities in the terrain which strongly influenced the area's development.

The sedimentary strata dip away from the Precambrian highland of northcentral Wisconsin, the Wisconsin Dome, as previously noted. Because the highest of those rocks, those that were on the dome, wore off first, the various layers are exposed as concentric bands curving around the dome, albeit asymmetrically and with some interruptions. The bands that are near the dome and were eroded relatively recently are roughly semicircular; but those farther from the dome, which have wider diameters and were eroded longer, are more irregular and incomplete in outline (plate 1 and figs. 5.1, 5.2).

Since some of the exposed strata are resistant and others weak, the terrain produced by weathering and erosion is one of long ridges or upland belts (the harder

Sediments are deposited, layer upon layer, over basement rock, such as the Precambrian shield margin, forming rocks of unequal resistance.

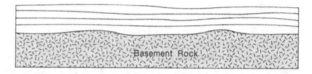

Basement rock is uplifted and sedimentary layers are bent. If erosion did not take place they would look like this.

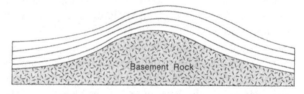

However, erosion takes place during and after uplift, so higher rocks are worn away, and the less resistant rocks erode faster than the resistant ones, causing an uneven surface.

Figure 5.1 Diagrams to illustrate how doming or arching of the basement rock tilts sedimentary rocks above. (Or subsidence of basins alongside can cause similar tilting of rock layers.)

layers) alternating with long lowlands (the weaker layers). Some ridges are continuous for scores of miles, and some are traceable well beyond the state's borders—for example, the Niagara formation. Others, cut by branching rivers, are greatly segmented and have had large sections removed (fig. 5.2).

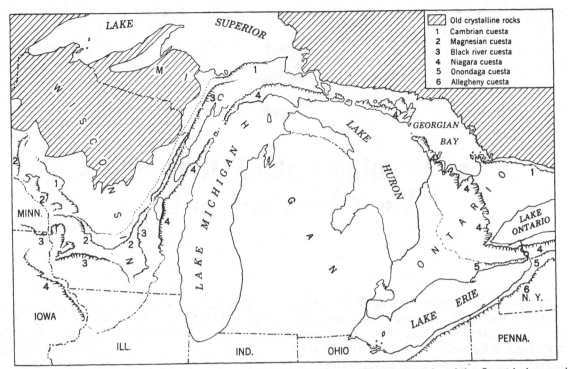

Figure 5.2 The Niagara Cuesta and other less prominent cuestas in the sedimentary rocks of the Great Lakes region. Parts of some ridges are deeply covered by glacial drift, and their approximate positions are shown by broken lines. Cuestas form concentric arcs around Wisconsin's northern highland. In Wisconsin the Cambrian cuesta has been called the Franconia or Ironton; the Magnesian is now called the Prairie du Chien; and the Black River is now called the Galena-Platteville. Notice how far east the cuesta of Niagara dolomite (4) extends, forming Niagara Falls between lakes Erie and Ontario. (McGraw-Hill Book Company. *Elements of Geography*, 5th ed., 1967, by Trewartha, et al.)

A ridge made in this manner by a slightly tilted stratum of hard rock is known as a *cuesta,* a Spanish term for a feature having this structure. In typical profile, one side of a cuesta is a more or less steep drop, and the opposite side is a long, gentle slope, usually miles in length. The steeper side, an escarpment, is where the eroded rock is exposed; it faces toward the high land (here, the Wisconsin Dome) from which the stratum is retreating. The gentler "dip slope" slanting away from the high land has the slant of, or is the slope of, the hard rock stratum itself, which is usually covered with weathered rock and soil (fig. 5.3).

A cuesta forms as relatively weak rock is removed from above a tilted hard rock layer and, more slowly, from under the hard layer's exposed edge. The hard layer's protruding edge is undercut, it cracks, and blocks of it fall off. In this way an escarpment maintains its steep or rough face and continues to wear back (fig. 5.3). A classic example of such a retreating escarpment is Niagara Falls. There the plunging Niagara River swirls powerfully at the fall's base and removes the weak shale beneath the hard, capping Niagara dolomite. Blocks of dolomite then tumble down, and the escarpment and waterfall slowly migrate upstream.

Wisconsin is just one of many regions of the world having cuestas formed in this manner—by stream erosion of gently inclined sedimentary strata of varying resistance. Wide plains that contain a series of alter-

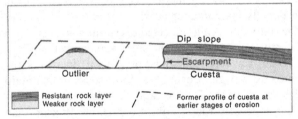

Figure 5.3 Profile of a retreating cuesta.

nating cuestas and lowlands are called *cuestaform plains.* In the United States they occur through much of the Midwest where strata have been warped into domes and basins, and also in such locations as the Gulf Coastal Plain, especially in Texas and Alabama.

In Wisconsin's cuestaform plain most of the ridges are protruding dolomite strata, and most of the lowlands are belts of shale and poorly cemented sandstone. (Sandstones can be extremely resistant if they contain a strong cementing material, and one of Wisconsin's sandstones—the Ironton—does produce a cuesta, but most sandstones in this area weather and erode rather easily, and so are not ridge-makers.)

In arid regions escarpments are typically bare, angular and precipitous; but in humid regions like Wisconsin they are more dissected and weathered, and usually have rounded slopes. In Wisconsin some parts of cuesta escarpments are bare rock cliffs, but a cuesta escarpment generally has the appearance of a ridge with

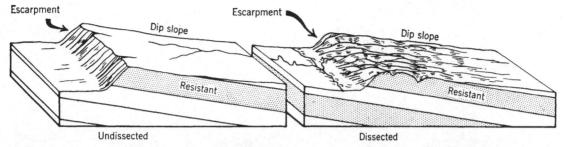

Figure 5.4 Diagram to illustrate form and structure of cuestas. Example at left is sharp and regular. Dissected form at right is more typical, especially in humid regions like Wisconsin. (McGraw-Hill Book Company. *Elements of Geography*, 5th ed., 1967, by Trewartha, et al.)

outcropping rock ledges that extends across the countryside, or of a belt of wasting hills. Where erosion has been slow to carry away fallen and slumped material, an escarpment may be smothering in its own talus and rock debris, or may be overgrown with trees or low vegetation, or may even be so gently sloping that it is cultivated (fig. 5.4).

In the broad view, then, we see broken arcs of escarpments facing northcentral Wisconsin from the southwest, south and east, indented or fragmented by streams crossing or gullying into them, and slowly backing downslope away from the dome. The outer escarpments have already receded beyond the state's borders. As time goes on the cuestas and intervening lowlands will keep retreating, and more and more of Wisconsin's Precambrian rock will reappear at the surface. Precambrian monadnocks are already poking through the thinning Paleozoic layers—the Baraboo Hills, Necedah Mound, the Waterloo outcrops, and other buried hills of ancient crystalline rock.

It should be remembered that most of the state's western sedimentary section bears no signs or only weak signs of glaciation; and that, in contrast, most of the eastern sedimentary section was glaciated recently and strongly. As one would expect, the cuestas stand out far more clearly in the Driftless Area. But in eastern Wisconsin they do show through the glacial drift in many places, and help form the framework of the terrain, so much so that geographers have called this section of the state "The Eastern Ridges and Lowlands" region (plate 4).

Across the Paleozoic section of southern Wisconsin, the western limit of glaciation runs north-south from central Portage County (east of Stevens Point) through western Waushara and southeastern Adams counties; by Wisconsin Dells, Baraboo and Prairie du Sac; west of Madison; and then on into Illinois. More will be said of glaciation in the following chapter.

Now we shall look at Wisconsin's individual cuestas and lowlands, starting on the Precambrian northern highland and moving outward (fig. 5.5).

The Cambrian Sandstone Lowland

As we leave the northern highland, the first lowland encountered (termed the "inner lowland") is the crescent-shaped Cambrian sandstone lowland mentioned in chapter 4, which roughly corresponds to the Central Plain on plate 4. In this lowland rocks from the Mount Simon Sandstone up through the Jordan Sandstone are exposed (plates 1 and 5).

The medium-grained to coarse-grained, weakly cemented Mount Simon Sandstone surfaces over a wide area. That earliest of Wisconsin's Paleozoic rock lies directly on the Precambrian shield. The succeeding thinly layered, shaly Eau Claire Sandstone forms rolling terrain of low relief with flat-topped benches or terraces, only occasionally making a ridge. In many road cuts and other exposures this fine-grained rock, sometimes containing thin dolomitic layers, may be seen overlying the cleaner, coarser, more uniform Mount Simon. These lowest formations, the first to cover the shield, are the last to be eroded from it.

Where rivers flow from the durable, recently uncovered Precambrian rock, dropping in still-unsmoothed channels to the weak sandstone lowland, rapids or cascades are common. During early settlement days, mills were built at these fall-line locations to utilize the waterpower to grind grain and saw timber; then other industries were established there, and around those sites grew towns, such as Chippewa Falls, Black River Falls, Wisconsin Rapids, and Waupaca.

In its western and central parts the crescent-shaped Cambrian sandstone lowland is broad, up to about 70 miles (110 km) across. There the combined strata are hundreds of feet thick, and they dip gently away from the dome. But the eastern part of the lowland is much narrower, because there the strata are thinner and dip at a steeper angle than in the west.

The Cambrian sandstone lowland is most distinctive in its middle section, known as the *Central Sand Plain* (compare plates 3 and 4). There it is exceptionally flat. In fact, this plain contains the state's largest single area of flat terrain. It covers all but the southwest corner of Juneau County, all but the southeast corner of Adams County, and parts of counties north-

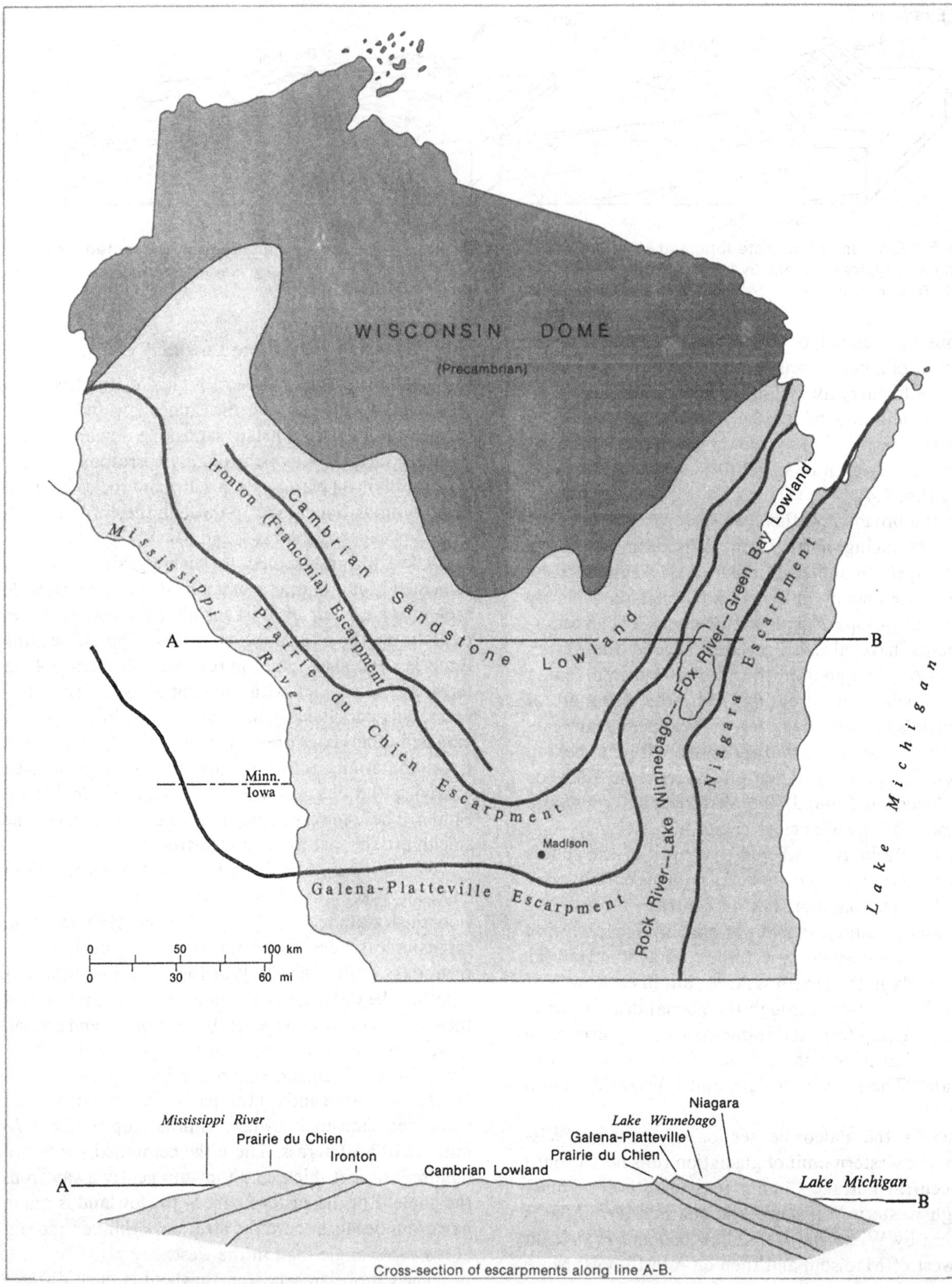
WISCONSIN DOME
(Precambrian)
Cambrian Sandstone Lowland
Ironton (Franconia) Escarpment
Mississippi River
Prairie du Chien Escarpment
Galena-Platteville Escarpment
Rock River—Lake Winnebago—Fox River—Green Bay Lowland
Niagara Escarpment
Lake Michigan
Madison
Minn.
Iowa
A
B
0 50 100 km
0 30 60 mi
Mississippi River
Prairie du Chien
Ironton
Cambrian Lowland
Niagara
Lake Winnebago
Galena-Platteville
Prairie du Chien
Lake Michigan
Cross-section of escarpments along line A-B.

Figure 5.5 Pattern of concentric cuestas in Wisconsin, generalized.

east, north, and northwest of them: northwestern Waushara, southwestern Portage, southern Wood, eastern Jackson, and northeastern Monroe.

The Central Sand Plain is bounded clearly on the east by a hilly belt (a moraine) that marks the limit of glaciation, and on the southwest and west by high and hillier terrain. On the north it grades into the shield's crystalline rocks under a cover of glacial drift. At the south near Wisconsin Dells it tapers to a point.

West of the level Central Sand Plain the Cambrian sandstone crescent is rough or hilly, because the Ordovician dolomite (which probably used to cover the whole crescent) has been removed only recently, and the land is in the process of being dissected and lowered. Still farther to the northwest, glacial deposition contributes to the unevenness. In its hilliness the western part of the sandstone crescent is topographically and visually much like Wisconsin's dissected Western Upland, with which it merges to the southwest.

The Central Sand Plain's levelness results not just from erosion planing down the sandstone. During the close of the glacial period much of the plain was covered by a lake whose bed of sediments was quite flat (plate 2). Additional deposits leveled the surface even more. Low places were filled with water-deposited materials which include alluvium spread by streams, sands and gravels laid by waters draining from the melting glacier, and sediments and organic matter that settled on the floors of the many marshes and other water bodies. Wind also spread sand and silt across the region.

The flatness of the Central Sand Plain causes drainage to be sluggish. Marshes and swamps abound in shallow depressions and along low-gradient watercourses. The largest marshy area—the largest in the state—is to the west, in southwestern Wood, eastern Jackson, northern Juneau, and northeastern Monroe counties; that is roughly within a triangle whose points are Black River Falls, Camp Douglas, and Wisconsin Rapids.

Near the exposed shield, where the sandstone is worn thin, subsurface drainage is impeded by the impermeable crystalline rock close below. This condition adds to the plain's swampiness. Castle Rock and Petenwell flowages are artificial reservoirs impounded by dams on the Wisconsin River, which flows across the heart of the Central Sand Plain. Many small streams are dammed by beavers, creating ponds and little lakes.

Much of the Central Sand Plain, as well as the rest of the Cambrian sandstone crescent curving through the center of the state, acts as an excellent groundwater reservoir. Generally precipitation is readily absorbed into the sandstone bedrock, and the water is easily recovered from it, but locally some sections have better water-intake and storage capability than others.

It is said that the most valuable part of sandstone is the part that is not there—the tiny, connected cavities among the sand grains. Those empty spaces permit the storage of copious amounts of groundwater and allow it to filter through the rock. Where gravelly and sandy drift of appreciable thickness is present, it too serves to store and filter water.

In the past the plain has been a region of low population density, with only a small proportion of its land developed. In its wide, unused expanses it was possible to establish the Necedah National Wildlife Refuge, the Central Wisconsin Conservation Area, the Sandhill Wildlife Area, Camp Williams Military Reservation, Volk Field, and the Black River State Forest. Also in this plains area is the Agricultural Experiment Station at Hancock in Waushara County, and near its western edge, the Winnebago Indian lands in eastern Jackson County. The towns of the plain are small, like Adams, Friendship, Necedah, Plainfield, Hancock, Babcock, and Wyeville. At or just beyond the margin of the plain there are larger cities: Wisconsin Dells, Mauston, Tomah, Black River Falls, Neillsville, and Wisconsin Rapids.

The Central Sand Plain is appreciated for its recreational value, and attracts many nature-lovers and sports enthusiasts. Picturesque craggy hills, described later, make this an especially scenic region. Other aspects of its increasing land use, including new agricultural developments, are discussed in the next chapter.

The Central Sand Plain can be viewed from many highways, though its more genuine, natural character is best observed away from the main roads. State Highway 13 from Wisconsin Rapids to Wisconsin Dells crosses the Central Sand Plain from north to south; and State Highway 21 crosses it from east to west between Coloma (Waushara County) and Tunnel City (Monroe County). There is a good overview of the plain and its craggy hills looking south from U.S. Highway 10 west of Neillsville (Clark County).

Stepping Up onto the Prairie du Chien Cuesta

Traveling northwest on Interstate 90–94 from Wisconsin Dells—by Lyndon Station, Mauston and Camp Douglas to Tomah—one observes a marked contrast between the flat sand plain to the northeast and the Western Upland rising to the southwest. The main divider between those two dissimilar topographic regions is the *Prairie du Chien Escarpment* (formerly called the Lower Magnesian Escarpment) which rims and overlooks the outer arc of the sandstone crescent. (It is so named geologically not because it is near Prairie du Chien, for it is not, but because that city is the type location of the Prairie du Chien Dolomite which forms it.)

This frayed escarpment is the eroding edge of the Prairie du Chien Dolomite (plate 1). It winds its way southwest from the Michigan border in Marinette County to south of the Baraboo Hills and the Lower Wisconsin River's big bend, and from there northwest to the "cheek" of the Indian Head. In the northwest the escarpment is obscured by drift, as it is east of the Central Sand Plain; but in the Driftless Area the escarpment is distinct along the plain's southwest border, and there one can plainly see the cuesta's margin, which is disintegrating and retreating. The many outliers it has left behind, scattered over the lowland, are reminders that the rocks of the cuesta once filled the space between those remnants and must have extended over the entire plain.

Ideally, the face of a cuesta consists of one main escarpment, but in some instances it consists of several steps. That is the case here. In simplified descriptions the Prairie du Chien Escarpment may be treated as one single rise, from the Cambrian lowland to the Ordovician (Prairie du Chien) upland to the southwest; but along much of its length the main escarpment is composed of two or three lesser escarpments rising in separate steps. The top level is indeed the hard Prairie du Chien Dolomite, but the lower steps are Cambrian.

In the spectrum of Cambrian formations, the first resistant one encountered as we progress outward from the shield is the Ironton Sandstone. The Ironton caps a cuesta in western Wisconsin that ranges in width from about 1 mile to more than 25 miles (40 km). It is the only sandstone formation in Wisconsin strong enough to form an escarpment. Its escarpment is considered the lowest step of the ascent to the larger Prairie du Chien Escarpment (figs. 5.6, 5.7). Though the Ironton Sandstone is a thin layer there, it is able to form a ridge because iron oxide and silica firmly cement its sand grains together. The rock layers above and below it are more easily eroded. In places it forms cliffs and crags, but most slopes formed by its outcrops are moderate (fig. 5.8).

Where the Ironton Sandstone forms an escarpment it stands as much as 200 feet (60 m) above the Central Sand Plain. This has long been called the "Franconia Escarpment" and is so labeled on many maps. However, that name is inappropriate since "Franconia" is no longer used as a rock name in geologic nomenclature. Because the escarpment is created by Ironton Sandstone it is fitting to call it the *Ironton Escarpment*.

If we think of Ironton Sandstone as the "frosting" of the Ironton Escarpment, then the Galesville Sandstone just below may be thought of as the "cake." And an eye-catching one it is—usually whitish in color, forming beautifully bright cliffs and exposures along the roads. It is seen in many outcrops along Interstate 90–94 from Wisconsin Dells to the northwest, and, farther east, in the walls of the Wisconsin Dells gorge.

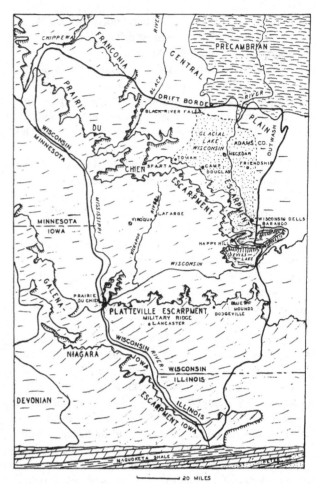

Figure 5.6 Escarpments of the Driftless Area as drawn by F. T. Thwaites, 1960. (Wisconsin Academy of Sciences, Arts and Letters)

The Galesville Sandstone has a wide range of hardness. Commonly it is so unconsolidated that it can be scratched with a stick, and weathers so rapidly that snowy piles of fallen sand envelop its base. But in some places it has become extremely hard. It is virtually quartzite in Silver Mound, which is 4 miles (6 km) northeast of Hixton in western Jackson County on the northwest side of Highway 95. That mound's upper hundred feet or so is cemented by silica, and is so strong and brittle that Indians used it to make tools and weapons (fig. 5.9). Numerous artifacts and chipped flakes have been found around the mound. Similar silicified sandstone is found in other hills in the vicinity, including the Neillsville mounds north of Neillsville.

The winding Ironton Escarpment extends from St. Croix Falls in the west to the Baraboo Hills, but it is not noticeable in glaciated eastern Wisconsin. It is the most striking escarpment in the Driftless Area, where it is recognizable as the first rise from the flat or gently rolling sand plain. On Interstate 90–94 from Wisconsin Dells to beyond Tomah, it is the irregular ridge of fairly even height roughly paralleling the highway on its southwest side, seen here and there from the

A PHYSIOGRAPHIC DIAGRAM
OF THE
DRIFTLESS HILL LANDS
BY
GUY-HAROLD SMITH

LEGEND

MARGIN OF THE WISCONSIN
DRIFT SHEET – – – – – – –

MARGIN OF THE EARLIER
DRIFT SHEET ·············

SCALE

0   10   20   30   40   50 MILES

Figure 5.7 A landform drawing of Wisconsin's Western Upland and parts of Minnesota, Iowa and Illinois. Cuestas are labeled. The Lower Magnesian Cuesta is now called the Prairie du Chien Cuesta. (*Annals* of the Association of American Geographers, vol. 1, 1941)

Figure 5.8 Ironton Sandstone used to be quarried from this hillside near the mouth of Poe Creek Valley in southeastern Monroe County. The unglaciated, driftless aspect is apparent. (Wisconsin Geological and Natural History Survey)

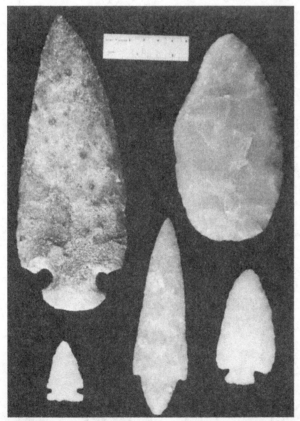

Figure 5.9 Projectile points and blades fashioned from "Hixton quartzite" (an archaeological term) from Silver Mound near Hixton, Jackson County. (By permission of the *Wisconsin Archaeologist*)

highway through the screen of trees. At some places, like Camp Douglas, it bends close to the highway (fig. 5.13). And at others it is seen as a distant row of dark hills. Sometimes one can differentiate a more distant ridge of hills beyond the Ironton; this may be the second step up the Prairie du Chien Escarpment or the highest surface of the Western Upland. Benches or terraces between the first-step Ironton Sandstone and the uppermost Prairie du Chien Dolomite are formed by dolomite or other relatively resistant strata.

U.S. Highway 12 and Interstate 94 rise over a northeast projection, or spur, of the Ironton Escarpment—from north of State Highway 21 (east of Tunnel City), on north to just over the Monroe-Jackson county line a few miles southeast of Millston, where the road drops to the lowland again.

Cuestas retreat in irregular fashion. The margin of the Prairie du Chien Cuesta curves in and out, and so accordingly do the lower steps of its escarpment, including the Ironton Escarpment. Eroding rivers have indented or breached the Prairie du Chien Escarpment and tributaries have further dissected the remaining divides, like many arms with fingers scraping up into the rock. As the tributaries reach each other across divides, like fingertips of two hands touching, parts of the cuesta's fringe become separated from the main, retreating escarpment and are left in front of the cuesta as isolated outliers. These may be individual hills or larger detached sections. In some places the dissected escarpment looks like an irregular belt of hills. But many of the outliers—near the escarpment, or tens of miles away from it—stand out dramatically as individual features, especially if they are on an exceptionally flat base like the Central Sand Plain.

Picturesque Outliers in the Sandstone Lowland

Outliers are probably the Central Sand Plain's most noted features. Locally they are called "mounds" or "bluffs." Many are more than 200 feet (60 m) high, and some more than 400 feet. They have attracted the photographer, sightseer and climber, and some are used as ski slopes. Many are the site of a park, wayside or campground, and have a trail to the top. From there one has a panoramic view across the plain, to other nearby mounds, and to the level Western Upland on the western horizon or to the glacial hills to the east. One can imagine the time, several thousand years ago, when a glacier-impounded lake covered the surrounding lowland and the highest mounds were islands in it. The lake waves beating against the Ironton Escarpment and its outliers helped steepen their sides (plate 2).

When first detached from a cuesta, an outlier may still be capped with the resistant protective rock layer that makes the cuesta. While that cover remains, the weaker rocks below are protected. On the outlier's vertical sides the exposed rock layers may acquire an in-

Figure 5.10 The Wisconsin Dells has been one of Wisconsin's leading tourist attractions for over a hundred years. Of the many interesting rock formations along the Wisconsin River there, the best known is Stand Rock, a pillar 46 feet (14 m) high, separated by erosion from the wall of the gorge. This picture was taken at the turn of the century. (H. H. Bennett Studio, Inc.)

and-out profile, either protruding or being indented according to their hardness or weakness (fig. 1.24). Sometimes the capping hard rock is so much more enduring than the lower rocks that it becomes a wide roof over a thin supporting pillar of worn rock. On a small scale, that is the way Stand Rock at the Wisconsin Dells was formed (fig. 5.10).

Once the protective cap is gone the softer rocks below it erode rapidly. Then the outlier's soft rocks erode along joints and weak beds, resulting in intricate, craggy shapes. Wind erosion—a kind of sand-blasting—plays a part in shaping and smoothing the rock into bizarre forms.

The oddly shaped hills and crags have inspired descriptive names. Several are called "Castle Rock" or "Castle Mound" because of their towers and buttress-like bases of sloping talus. Other popular names used for a number of mounds are "Steamship" (because of

the resemblance) and "Wildcat" (because wildcats lived on them).

Some outliers have lasted an unusually long time because exposure strengthened the sandstone by case-hardening, a process by which the surface of porous rock is coated by a cement or crust produced by evaporation of mineral-bearing solutions. But weathering and erosion continue relentlessly. Loosened material falls and keeps collecting around an outlier, so in time it acquires a conical shape. Gradually the outlier becomes lower and wider, and ultimately merges into the general level of the plain, adding its sand (its most enduring ingredient) to the overall sandy surface (fig. 5.11). Rollin Salisbury and Wallace Atwood described the process well in 1900: "Rains, winds, frost, and roots are still working to compass the destruction of these picturesque hills, and the talus of sand bordering the 'castle' is a reminder of the fate which awaits them."[1]

Intricately etched outliers, which today are characteristic features of the unglaciated part of the sand plain, must have existed also in the glaciated part of the sandy crescent before the scraping glacier toppled them or ground them down. Only a few outliers remain in the glaciated section, near its margin where the ice was too thin and weak to completely destroy them. Some of these have glacial drift on top, indicating that the ice overrode them. Others were glaciated only at their base by thin ice.

Because of the great number of named outliers in the sand plain, only a few that are well known and easily seen will be mentioned. Interstate 90–94 and U.S. Highway 12 pass among many of them, in various stages of erosion, along the Ironton Escarpment.

At the town of Camp Douglas is a cluster of outliers, including Target Bluff just southwest of the highways, smaller Castle Rock (also called Battleship Rock) just northeast of them, and Chinaman Bluff about a mile northwest of Camp Douglas on the northeast side of the interstate highway (figs. 5.12, 5.13, 5.14).

Figure 5.11 A sketch of features of Cambrian sandstone, published in 1862, made "to show the irregular mode in which this sandstone weathers, and the curious conical and mound-like forms which it occasionally assumes." (Hall and Whitney, 1862)

1. Rollin D. Salisbury and Wallace W. Atwood, *The Geography of the Region about Devil's Lake and the Dalles of the Wisconsin* (Madison: The State of Wisconsin, 1900), Wisconsin Geological and Natural History Survey, Bull. no. 5, p. 71.

Figure 5.12 Looking east over the Central Sand Plain from an airplane above the Ironton Escarpment. The town of Camp Douglas is at the right, and Volk Field and Camp Williams at the left. (Compare with the southeast corner of the fig. 5.13 map.) Notice how transportation routes converge here. Target Bluff (right of center) stands to the left (northeast) of a railroad line, between it and U.S. Highway 12 which curves around it. Across double-highway Interstate 90-94 is smaller Castle Rock. Another large outlier is to its left. (Carl Guell)

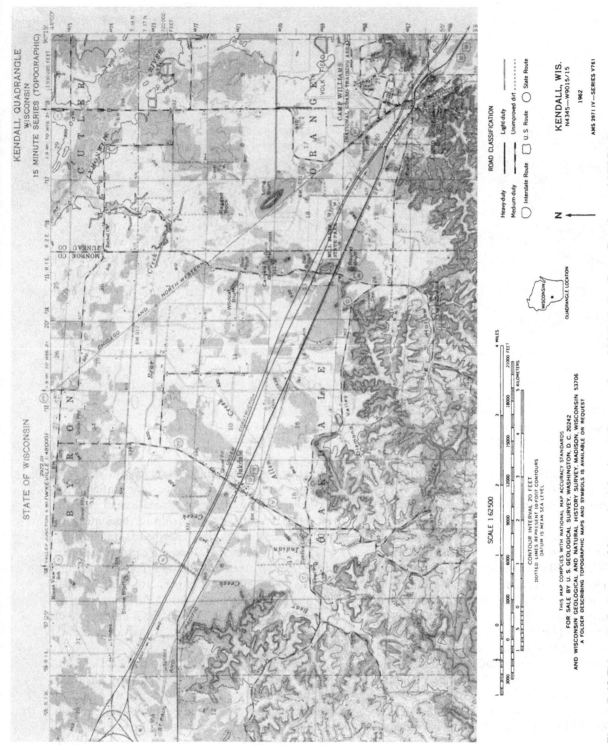

Figure 5.13 This part of a topographic quadrangle shows the mounds and bluffs in the Camp Douglas and Mill Bluff areas along Interstate highways 90–94 and U.S. highways 12 and 16. The roads lie along the foot of the Ironton Escarpment. Contour lines connect points of equal elevation above sea level. The more closely spaced the lines are, the steeper the slope. The area around the town of Camp Douglas can be related to the aerial photograph, fig. 5.12. (U.S. Geological Survey)

Figure 5.14 Castle Rock, across the highways from Camp Douglas, as it looked in 1934 before modern highways were built. Its lower part is Galesville Sandstone, its upper part Ironton Sandstone. (Wisconsin Department of Natural Resources)

About 3 miles northwest of Camp Douglas is Mill Bluff in the state park of that name. It stands at the eastern border of Monroe county between the interstate highway and U.S. Highway 12. A trail leads to the top, and from there one can look out over the plain and see other outliers—Bee Bluff, Camels Bluff, Bear Bluff, and Wildcat Bluff to the north; Long Bluff and Ragged Rock to the northeast; and Round Bluff and others to the south. The highest outliers in this cluster, as in the Camp Douglas cluster, rise about 200 feet (60 m) above the plain (figs. 5.13, and 5.15 through 5.19).

Clusters of outliers are seen also in the Tomah-Wyeville-Valley Junction crossroads area. Well-known mounds in the western section of the sand plain are Bruce Mound east of Merrillan, and Saddle Mound south of Pray in eastern Jackson County, both of which rise more than 350 feet (about 100 m) above the lowland.

Petenwell Rock (or Peak) is a landmark along the west side of the Wisconsin River, on the south side of State Highway 21 east of Necedah. North of it is Petenwell Flowage. This outlier is prominent because it looms up 222 feet (68 m) from the water. Its elevation is 1,100 feet (338 m). The upper part is a narrow, serrated ridge of light, friable sandstone, and at its base is sandy talus.

In central Adams County, State Highway 13 runs just east of several interesting sandstone outliers.

One is Roche à Cri (pronounced Roash a Cree), about 2 miles north of Friendship in the state park bearing its name. It has been described as standing like a fragment of a great wall. Its south end is a precipice of more than 200 feet, and its top is flat, without pinnacles. Its base, which is elongated north-south, is about 1,300 feet (400 m) in length (fig. 5.20).

Relatively small Rabbit Rock lies 3 miles north of Roche à Cri right along the road, with a wayside park next to it.

Friendship Mound is south of Roche à Cri and immediately north of Friendship. Its high south face overlooks the north edge of that town; and its southeast corner rises sharply from the highway along Friendship Lake. Capped with Ironton and Lone Rock sandstone, Friendship Mound is higher and more rounded than its northern neighbor, and is more than three-quarters of a mile long.

Rattlesnake Mound, another sandstone outlier, is 5 miles south of Friendship Mound and almost a mile west of State Highway 13, which it parallels. It is narrow, and one mile and three-quarters (2.8 km) long, divided into two parts by a central depression. About 2 miles to the west is Quincy Bluff, similar in shape.

Figure 5.15 Mill Bluff, as seen looking north from Round Bluff, with other bluffs in the background on the flat Central Sand Plain and old floor of Glacial Lake Wisconsin. (Robert F. Black)

Figure 5.16 Hard rock layers, responsible for Mill Bluff's durability, jut out over softer, faster eroding sandstone. (Wisconsin Department of Natural Resources. Photographer, Robert Espeseth)

Northwest of the Central Sand Plain, but still within the sandstone crescent, is Wildcat Mound, about 4 miles east of Humbird in southwestern Clark County. This mound is noted for its distinctively variegated, buff-colored sandstone, which is stained and cemented by iron oxide and is locally called "zebra rock" because of its coloration. A path up the mound leads to good exposures of the streaked and mottled rock. This mound is not to be confused with Wildcat Bluff, mentioned previously, or Wildcat Mountain in Wildcat Mountain State Park in the Western Upland (eastern Vernon County).

Another landform feature of Cambrian sandstone should be mentioned here, although it is on the outer margin of the sand plain, south of the Baraboo Hills. That is the natural bridge about a mile northeast of Leland, northwest of County Highway C, in Natural Bridge State Park (fig. 5.21). This bridge of Mazomanie Sandstone is about 40 feet (12 m) high, and the opening beneath the arch is about 35 feet wide and 25 feet high. At the base of the bridge is a rock shelter in which prehistoric Indians lived when they sought protection from the weather. Artifacts of stone and bone found there date back nearly 7,000 years. This is Wisconsin's largest natural bridge, and one of the largest

Figure 5.17 From atop Mill Bluff, looking north, one sees a number of mounds, including Bee Bluff, the nearest, and Ragged Rock in the dim distance at the right. These outliers, like many in the Central Sand Plain, were islands in Glacial Lake Wisconsin during the Ice Age. (Wisconsin Department of Natural Resources. Photographer, Dean Tvedt)

a

b

Figure 5.18 (a) Ragged Rock, in a mostly wooded section of the Central Sand Plain. **(b)** A close look at the top of Ragged Rock with its craggy pinnacles, features commonly found on wasting hills in the Driftless Area. (George Knudsen, Wisconsin Department of Natural Resources)

Figure 5.19 In the Mill Bluff area these petroglyphs (carvings in rock) were scratched in soft sandstone late in prehistoric time. Petroglyphs are found elsewhere in Wisconsin too. (George Knudsen, Wisconsin Department of Natural Resources)

Figure 5.20 Roche à Cri in central Adams County, seen from the south-southeast. The flat surface around this outlier was formerly the bed of Glacial Lake Wisconsin. (Lawrence G. Monthey)

Figure 5.21 The Natural Bridge north of Leland, southcentral Sauk County, as photographed by F. T. Thwaites in 1923. The opening, or rock shelter, under the bridge was inhabited by prehistoric Indians in times of bad weather. (Wisconsin Geological and Natural History Survey)

Figure 5.22 A mesa of Mazomanie Sandstone northeast of Leland about 1920. (Wisconsin Geological and Natural History Survey)

in the Midwest. A smaller natural bridge of sandstone arches over a stream at Rockbridge, about 10 miles north of Richland Center.

Other Views of the Central Sand Plain

Many flat-topped outliers of the Central Sand Plain resemble the mesas and buttes of horizontal sedimentary rock found in arid regions. ("Mesa" is the Spanish word for "table." Buttes, like mesas, are angular or table-shaped, but smaller. See fig. 5.22.)

Lawrence Martin, remarking about the resemblance in 1916 in his book *The Physical Geography of Wisconsin,* pointed out that in central Wisconsin one sees "landscape features totally unlike those anywhere else in the United States east of the Mississippi River. . . . They have the straight lines, steep cliffs, and sharp angles of an arid country rather than the soft curves of a humid region. This is the very frontier of the true West."[2]

Contributing to the misleading arid appearance of this landscape are accumulations of sand, and the prickly-pear cactus, which grows in places where moisture drains quickly from the porous sand.

Of part of Jackson County, Martin also wrote: "One might perfectly well be in Wyoming rather than Wisconsin. The scrub timber grows in bunchy groups. There is no grass, and the white sand appears between the shrubs. The hill slopes disclose vertical cliffs, angular profiles, flat-topped ridges, teepee buttes, much as on part of the Great Plains." But the similarity is incomplete and illusory. Nearby, as he points out, are "swampy plains, which need ditches to drain away the water" and "the homelike aspect that goes with waving fields of grain, prosperous farm houses, and thriving villages."[3]

Martin gives special attention to features that could be seen along railroads, because the train was the common mode of cross-country travel when he wrote, and people traveling through the state would most likely see those places. He also remarked about scenic features along main highways and at their crossings.

As one would expect, the first roads and railroads were built, insofar as possible, over flat land. So they were laid along the foot of the Ironton Escarpment rather than in the dissected, hilly upland just to the southwest, where many bridges, much grading, and an occasional tunnel would be required. Interstate Highway 94 to Minneapolis-St. Paul was constructed along the same convenient route. From that transportation corridor main transverse routes crossing western Wisconsin make use of river valleys cut through the upland.

Before the era of roads and railroads, prominent mounds and crags were landmarks and convenient meeting places for Indians and for other people who came later. High ones were visible from a great distance. Travelers navigated from one to another, and from the tops of those mounds fire and smoke signals could be seen far away. Mounds near rivers must have been regular places to meet since water was the main artery of travel in early days. It is not coincidence, then, that transportation routes of today, many of which were built on old Indian trails, pass by some of the more spectacular mounds and crisscross there. Centuries ago some of those outliers that we see now were larger and differently shaped, of course; and those that we consider large today may be reduced or changed in appearance in time to come. Then people may wonder why they have the special names they have or why they were thought to be outstanding.

The Prairie du Chien Cuesta of the Western Upland

The irregular escarpment of the Prairie du Chien Cuesta runs diagonally from northwest to southeast through the Driftless Area to the Lower Wisconsin River (figs. 5.6, 5.7, and plate 4). Most of the cuesta's dip slope in the unglaciated Western Upland is greatly dissected, but Prairie du Chien Dolomite still surfaces the hills north of the river and thus controls the topography, which is a product of stream erosion. Streams have carved most of the land into narrow ridges separated by deep, steep-sided valleys that branch out in a dendritic, or treelike, pattern. The dolomite makes a fertile soil on the ridgetops. When washed into the valleys it helps to fertilize the more barren soil there, which is derived from Cambrian sandstone.

Like the Cambrian sandstone, Prairie du Chien Dolomite has a broader area of exposure in western Wisconsin than in eastern Wisconsin, and for the same reasons—greater thickness and a gentler dip of the strata in the west. Also, the absence of drift in the unglaciated area allows it to show at the surface.

The levelest, least-dissected part of the Prairie du Chien Cuesta in Wisconsin is that section farthest to the northwest, the Indian Head's "jaw" in Pierce, St. Croix and southern Polk counties (plates 1 and 4). There the cuesta surface dips gently toward the west and southwest, and the upland is a plain or a relatively flat, low plateau, leveled in part by glacial filling. The north and northwest sides of the "jaw" section are heavily covered with drift. On the east side the irregular escarpment trends north-south to central Pepin County at the Chippewa River. The escarpment is quite evident where U.S. Highway 12 and Interstate 94 cross it near Knapp (Dunn County). Near the Chippewa, Mississippi and St. Croix rivers are hilly belts such as commonly develop where tributaries of large rivers are vigorously dissecting adjacent uplands. Overlying part of the Indian Head's "chin" are outlier hills and ridges of younger St. Peter Sandstone and Sinnipee Dolomite.

From the Chippewa River southeast to the La Crosse River, in the Driftless Area, dissection has progressed further. Practically no level upland remains; the area is a maze of stream-eroded ridges and valleys. Hilltops rise to 1,100 and 1,300 feet (400 m) above sea level, and many streams have valleys 400 or more feet deep. This section includes all of Buffalo and Trempealeau counties, and parts of La Crosse, Monroe, Jackson, Eau Claire, and Pepin counties.

Down-cutting and valley-widening by the Mississippi and its large tributaries—the Chippewa, Buffalo, Trempealeau, Black, and La Crosse rivers—have effectively broken up the Prairie du Chien Dolomite cover.

2. Martin, p. 299 (1932 ed., p. 317).

3. Ibid., pp. 302, 303 (1932 ed., pp. 319–20).

Figure 5.23 Hills of the Driftless Area near Westby in northcentral Vernon County. Roads follow the valleys. Farms are on the flatter land of valley floors and wide hilltops, while steep slopes are left in trees. Contour strip-cropping helps control erosion. (Carl Guell)

Much of it has been removed, and wide areas of Cambrian sandstone are exposed. Interstream ridges are narrow. Those still capped with dolomite are craggy, whereas those without the dolomite cap are lower and more rounded. Main valley floors are rather flat and generally a mile or more wide.

Farther south in the Driftless Area, between the La Crosse and Wisconsin rivers, the cuesta is more intact, but highly dissected nevertheless (fig. 5.23). In this hilly region the areas of ridge top and valley bottom are about equal. Most ridge crests are only a fraction of a mile wide. Many have an elevation of more than 1,200 feet (365 m) above sea level; not far south of Interstate 90 some have elevations of more than 1,300 or 1,400 feet. Main roads tend to follow valleys, going to the ridge tops where they are broad enough or have to be crossed. Some ridges are wide and unbroken for long distances. One ridge west of the Kickapoo River, extending from southwestern Monroe County to Prairie du Chien, is more than 50 miles (80 km) long (plate 1). Cashton, Westby and Viroqua are situated on it,

and State Highway 27 was built along it. This ridge is topped along most of its length with St. Peter Sandstone, and Sinnipee Dolomite overlies the sandstone from Seneca south. Otherwise, most ridges in this region are capped with Prairie du Chien Dolomite.

Here, as in much of the Western Upland, the high percentage of land in slope creates problems. Floods can occur when runoff is rapid and concentrated in narrow valleys; and in many valleys they are expected every spring at least. The threat of rapid erosion requires the use of conservation measures to preserve soil and retard gullying on cultivated and grazed slopes. Steep slopes are generally left wooded.

In the area of Coon Valley, in northcentral Vernon County, the nation's first watershed project was developed in the 1930s by the U.S. Soil Conservation Service, the University of Wisconsin, and local farmers. Its aim was to control erosion on 92,000 acres, on 750 farms, in La Crosse, Vernon and Monroe counties. It demonstrated the effectiveness of soil and water conservation practices such as strip-cropping, contour

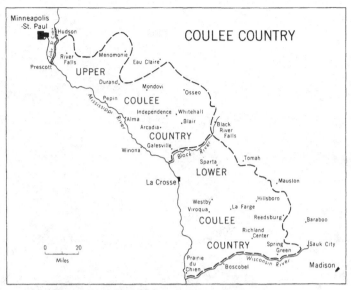

Figure 5.24 The region known as "coulee country" in westcentral Wisconsin. (Cotton Mather)

plowing, terracing, tree-planting, gully control, and controlled grazing. This system of land management became prevalent in the Midwest, and the principles of soil and water conservation that proved successful here were adopted across the country and around the world.

The hilly land east of the Mississippi River from southern St. Croix County to the Wisconsin River is "Coulee Country" (fig. 5.24). The region best known as coulee country, however, is that part from the Wisconsin River Valley north to the Trempealeau River, especially the area around La Crosse.

Coulee country does not extend northward into the heavily glaciated area. It is driftless, except for its northernmost part (from Buffalo County north), and that has only a thin covering of old drift, not enough to have significantly modified the valleys or altered their surface drainage.

Drainage is a key element in the definition of a coulee. The word "coulee" comes from the French "couler" which means "to flow." Generally it means a ravine or deep gully, but it has different specific meanings in different regions. In westcentral Wisconsin, a coulee is a narrow, steep-sided valley that is closed at the upper end and has a rather flat, sandy or silty floor, and that is drained by a small stream, which may flow only in spring or also after rains, or continuously in some cases. Typical coulees, it is said, are dry except in spring. Valleys of large prominent streams are not coulees. Coulees are tributary to a large permanent stream or to another coulee. There are many such small valleys in the intricately dissected hill region (figs. 5.7, 5.25).

In spring, when snow is melting and rains are heavy, water flows down the steep walls of the coulees and spreads its sediment, or wash, on the valley floors. Later many of these streams dry up and remain dry for the rest of the year because the less-abundant precipitation can readily drain through the porous bedrock, and because the small water-collecting area of a valley does not provide enough runoff to maintain a permanent stream.

Roads running up coulees are generally dead-end ones, for a coulee tapers off into a narrow ridge or other divide. Many of these separated, secluded protected valleys are named for families who live or have lived in them, like Jostad Coulee, Olson Coulee and Hagenbarth Coulee. Others are given applicable names like Chipmunk Coulee, German Coulee and Sweden Coulee. Many Norwegians settled in this region. It was as much like the land they left as any place hereabout: deeply dissected terrain with narrow, private valleys.

South of La Crosse, coulee-type valleys are commonly called "hollows," the English counterpart of the French word. Occasionally they are called "bottoms." Few typical coulees are found south of the Wisconsin River.

The Lower Wisconsin River makes a major cut through the Prairie du Chien Cuesta where it flows west. Its valley floor is in Cambrian sandstone except just near its mouth. There, as it approaches the Mississippi, the river intersects the hard, southwest-dipping Prairie du Chien Dolomite where it is thick. It cuts through the dolomite more slowly than it does through the sandstone, so the valley is narrower near the mouth than it is upstream—a reversal of the usual valley shape, as river valleys normally widen toward their mouth. The Wisconsin River Valley is 4 miles (6 km) wide in the weak sandstone near Sauk City, but only a half-mile wide at Bridgeport near the Mississippi River.

In bluffs of the Mississippi River Valley in this area the strong Prairie du Chien Dolomite has created cliffs and precipices that add much to the scenery. This rock retains most of its original thickness toward the south, where it was buried longest (fig. 5.26). The dip of the formation can be followed in outcrops. Its full thickness is revealed several miles north of Prairie du Chien. Farther north along the river, the formation is seen to occupy a progressively higher position on the bluffs, until near the Black River it occurs only thinly in outliers—its lowest beds cover the highest elevations; its upper beds are completely eroded away.

The Prairie du Chien Cuesta in Central and Eastern Wisconsin

As we follow the Prairie du Chien Cuesta eastward into Sauk County, we find that its retreating margin still covers the west end of the Baraboo Hills. Still farther east, in Dane and Columbia counties, the Prairie du Chien Escarpment has a relief of about 300 feet (90 m). One is aware of ascending from the Cambrian sandstone lowland onto the cuesta when going

a

b

Figure 5.25 (a) Looking toward the head of a coulee. Coon Creek southeast of La Crosse. (David H. Thompson, Environmental Images) **(b)** A coulee in Trempealeau County—looking from a ridge at its head down toward its mouth. (Gwen Schultz)

Figure 5.26 A bluff of Prairie du Chien Dolomite along the Mississippi River near Wyalusing, Grant County. (Gwen Schultz)

south from Lodi to Dane, or from Poynette to the level Arlington Prairie around Arlington—traveling among the escarpment's outliers and up its dissected, hilly face. The escarpment and its outliers have been given rounded contours by glaciation and a cover of vegetation. The escarpment is responsible for the steep grade on U.S. Highway 12 at Springfield Hill, about midway between Madison's west side and Sauk City.

Before glaciation, streams eroded through the Prairie du Chien Cuesta, exposing Cambrian sandstone in central Dane County, and along the wide valley of the Yahara River (formerly called the Catfish River) to the Rock River, and in the Rock River Valley from Jefferson south to Janesville. Now this break in the cuesta is heavily drift-covered (plate 1).

The dip slope of the Prairie du Chien Cuesta narrows as it continues on to the northeast through Columbia County. There the escarpment is not apparent in the landscape because its crest has been abraded by glaciers, and the drift that coated it has more or less filled in the lowland at its base. Though this eastern portion of the escarpment is smoother than the unglaciated, highly dissected western portion, it does have indentations and salients characteristic of a retreating escarpment. The scouring ice sheets that ground off its irregularities also removed many vulnerable outliers. Those that remain have had their fragile crags and pinnacles scraped off. They seem to be molded into the general terrain and do not stand out boldly like those in the Central Sand Plain.

Intervening Softer Rocks

In our traverse outward from Wisconsin's Precambrian northern dome across alternating lowlands and cuestas, we have so far crossed one of each—the crescent "inner lowland" of Cambrian sandstone, and the broken, rimming Prairie du Chien Dolomite cuesta. As we continue outward we find, overlying the Prairie du Chien Cuesta, a layered group of softer strata whose main formation is St. Peter Sandstone. Where these soft

rocks form a lowland it is not as wide as that formed by the Cambrian rocks. These are not as thick, and are more easily worn away. Some of the St. Peter Sandstone is so weakly held together that grains fall loose at a light touch, but some that is naturally cemented is more lasting.

In eastern Wisconsin the soft St. Peter and overlying transitional rocks have been eroded back, right up to, or within a few miles of, the next escarpment, that of the strong overlying Sinnipee Dolomite (plate 1). They are largely covered by drift. In western Wisconsin, as noted before, sizable outliers of St. Peter Sandstone remain on the "chin" of the Indian Head (St. Croix and Pierce counties), and atop the long north-south ridge through Vernon and Crawford counties, and in other scattered patches. White or light buff exposures of the St. Peter enhance cliffsides, including those along the Mississippi River in southern Wisconsin.

A small but noteworthy St. Peter Sandstone outlier in southcentral Wisconsin is that on Gibraltar Rock, a bluff south of the lower end of Lake Wisconsin, southwest of Okee (fig. 5.27). Gibraltar Rock is, in fact, an outlier of both Prairie du Chien Dolomite and St. Peter Sandstone: Cambrian sandstone forms its lowest part; that is overlain by Prairie du Chien Dolomite; and surmounting the dolomite is St. Peter Sandstone, which forms a cliff at the top. Glacier ice rounded Gibraltar Rock's east side and left deposits on its top. This bluff, the highest outlier in its area, rises more than 300 feet above the lake level to an altitude of 1,247 feet (380 m). A trail winds up its wooded north side.

The Galena-Platteville Cuesta and Its Significance in Southwestern Wisconsin

In distance away from the Precambrian dome of northern Wisconsin, the next major cuesta is the Galena-Platteville Cuesta, formed by the Sinnipee dolomitic rocks—the Platteville, Decorah and Galena formations (plate 5 and figs. 5.5, 5.6 and 5.7).

The escarpment of this cuesta, the Galena-Platteville Escarpment, faces the northern dome as the other escarpments do. It enters the state in the northeast in southeastern Marinette County and trends southward roughly parallel to the Prairie du Chien Escarpment, lying several miles east of it (plate 1). Since both escarpments are sinuous, the distance separating them varies, from less than 2 miles to more than 10 miles.

Farther south, the Galena-Platteville Escarpment lies west of Lake Winnebago, trending southwest. In Columbia and Dane counties it loses its identity, having been cut through and fragmented by stream erosion. The Madison area's Four Lakes lowland and the Rock River valley create much of that interruption. Drift blankets the area, as it does the whole eastern part of the cuesta thus far described, so one would not be aware

Figure 5.27 Gibraltar Rock south of Lake Wisconsin, across the Wisconsin River from Merrimac. (Madison Metropolitan School District. Photographer, Ron Austin)

of the escarpment's outline and extent unless one knew the bedrock structure under the drift.

In driftless western Dane County, however, the escarpment takes on a continuous, obvious form, and from there west for more than 70 miles (110 km), along the south side of the Lower Wisconsin River through central Iowa County and Grant County, it is one of the state's most conspicuous escarpments. An outlier of the Galena-Platteville Cuesta juts north of the Wisconsin River into Crawford County west of the Kickapoo River, where it overlies the southern third of the north-south outlier ridge of St. Peter Sandstone (fig. 5.6 and plates 1 and 5).

At the state's western border, the Galena-Platteville Cuesta makes steep bluffs and cliffs where the Mississippi River has cut a trench through it. West of the Mississippi its escarpment reappears, curving northwest across Iowa and into Minnesota.

South of the Lower Wisconsin River, the Galena-Platteville Escarpment is quite imposing. It is dissected by short, steep-gradient streams and consists of wooded bluffs, deep tributary valleys and ravines. If it were not for the Wisconsin River at its base, the north-facing escarpment here would be only 100 to 200 feet high. But the river eroded down through the underlying St. Peter Sandstone and Prairie du Chien Dolomite into

the Cambrian sandstone, carving a valley as much as 500 feet (150 m) deep. Later the valley was partly filled with 100 to 200 feet of sands and gravels flushed from the melting glacier. Farming is done on much of the level and nearly level land at the foot of the escarpment. That level land is partly the river's floodplain, but it is mainly terraces or benches of glacial outwash which are somewhat higher and better drained than the floodplain.

The cuesta's south-facing, longer dip-slope side is smoother—level to gently undulating, with some hilly areas along the rivers (fig. 5.7). Parts of it are dissected as much as the hill land north of the Wisconsin River, but the proportion of level land in the south is much greater.

The Galena-Platteville Cuesta constitutes the upland of southwestern Wisconsin. It slopes gradually to the south into Illinois with the dip of the dolomite. Summits are of a rather uniform elevation, but standing conspicuously higher are a number of mounds—outliers of the next, outer cuesta, the Niagara Cuesta. Valleys of the Galena-Platteville Cuesta generally remain quite narrow because streams are cutting through two resistant dolomite groups, the Galena-Platteville and the Prairie du Chien. The Prairie du Chien and St. Peter formations crop out along the south-flowing rivers and the main tributaries which are dissecting the cuesta. These rivers are, from west to east, the Grant and the Platte (both tributary to the Mississippi), and the Pecatonica and the Sugar (both tributary to the Rock).

The still-unbreached divide between the cuesta's short north-flowing and longer south-flowing streams is one of Wisconsin's best known topographic features—*Military Ridge*. It is the fairly level stretch of dolomitic upland aligned east-west just south of the Galena-Platteville Escarpment, generally about 10 or 12 miles south of the Wisconsin River Valley. Its elevation is about 1,200 feet (365 m) above sea level (figs. 5.6, 5.28).

When settlers came to southwestern Wisconsin it was seen that this long, even, well-drained ridge was the best route along which to construct a road. So along its crest in the 1830s the Military Road was built. That road, which gave its name to the ridge, connected Fort Howard at Green Bay, Fort Winnebago at Portage, and Fort Crawford at Prairie du Chien. Later a railroad too was built along this ridge. No stream of significant size had to be bridged for more than 60 miles (100 km). The old Military Road in this ridge section has become present U.S. Highway 18, from Mount Horeb past Blue Mound (an outstanding landmark to early travelers) and through or near towns strung along the ridge, including Barneveld, Ridgeway, Dodgeville, Cobb, Montfort, Fennimore, and Mount Hope (fig. 5.28). At Bridgeport, where the river trench is narrow, Highway 18 descends the escarpment and crosses the Wisconsin River to reach Prairie du Chien.

Lowlands along the Wisconsin River also were a convenient early route of travel through the hilly Western Upland, as was the river itself. Railroad and highway traffic made use of this level corridor.

The confluence of the Wisconsin and Mississippi rivers was a strategic site fought over by the French and English when they were vying for control of this region's fur trade. One of the most successful early fur-trading posts was Prairie du Chien, a few miles north of the confluence. This site, where river routes converged, was also a natural place for canoeists to meet in frontier days and long before. Here the scenery was the most magnificent for miles around: the dolomite-capped bluffs lining the rivers; the terraces along the mighty Mississippi, which Indians called "The Father of Waters"; the meanders and islands on the floodplain. Because the Driftless Area lacked lakes and swamps, the large rivers and their sloughs and backwaters had a special appeal. From high bluffs there were panoramic vistas and one could view the rivers joining 500 feet below. It is known that Indians favored this place because many of their effigy mounds, trails, village sites, and planting grounds have been found there. Birds are abundant along the Mississippi River Valley, as it is a major migratory route for them. Long ago it was predicted that that scenic, historic river junction would become a tourist center, and so it has. In 1917 Wyalusing State Park was established there on the south side of the Wisconsin River.

The southwestern upland was the first part of Wisconsin to be settled as the west-moving tide of population came to the Midwest. Belmont in western Lafayette County became the Wisconsin Territory's first capital in 1836. The Galena-Platteville Cuesta's dip slope was one of the state's choicest agricultural regions. Over much of it soils were fertile as a result of a combination of conditions: they developed from dolomitic rock; and on the broad divides the native vegetation was prairie grass, the roots of which add much humus to the soil; and the upland, like much of the Driftless Area, was (and is) coated with fine, fertile dust (loess) blown there by prevailing westerly winds.

Those who would farm and build homes found the oak openings of southwestern Wisconsin inviting. They were expanses of tall prairie grasses and wild flowers with oak tress scattered throughout, having the aspect of a natural park or orchard. Some of those oak trees still survive. In the oak openings the soils were productive, well-drained and easy to till. Though the sod was hard to break up, it was not necessary to remove boulders as in the glaciated area, or to hew down a whole forest and dig up its stumps. Still, enough of the area was wooded so that firewood and building material were available.

But desirable agricultural land was not the main reason this section of the state was the first to be settled and developed. The primary, stronger attraction was lead.

The Lead-Zinc District of Southwestern Wisconsin

Southwestern Wisconsin makes up the major part of the Upper Mississippi Valley lead-and-zinc district, which also includes northwestern Illinois and a fringe of northeastern Iowa (fig. 5.29). Its "glory days" of mining are past, but reserves remain, and what the future may be is uncertain. Within this district there once were thousands of lead mines and hundreds of zinc mines. The mines were relatively small because deposits were small and dispersed, but the total yield was considerable and made this are one of the world's foremost suppliers of lead and zinc during its peak production periods. More than 1.2 million tons of zinc and nearly 100,000 tons of lead were mined in southwestern Wisconsin from 1910 to 1974. For earlier years accurate production figures are not available, but it is estimated that some 250,000 tons of zinc and 350,000 to 400,000 tons of lead were produced there in that period following 1800. Small amounts of copper have been commercially mined in southwestern Wisconsin in years past.

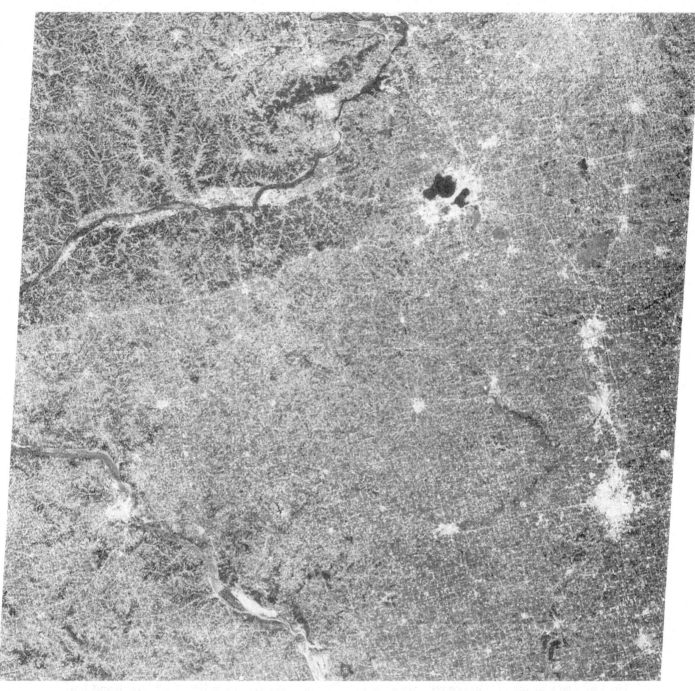

Figure 5.28 Satellite photo centered on southwestern Wisconsin from 570 miles (917 km) above the earth. Madison surrounds lakes Mendota and Monona, which appear dark. The large city in the lower right is Rockford, Illinois. North of it is Beloit at the Wisconsin-Illinois border, and north of it is Janesville. The Mississippi River curves across the lower left. In the upper left the Wisconsin River bends around the Baraboo Hills and flows west to meet the Mississippi. U.S. Highway 18 follows the crest of Military Ridge west from Madison. South of it is the Galena-Platteville Cuesta's dip slope. The field pattern on the relatively level area there and extending into Illinois is more even than that of the rough escarpment on the north side of the highway and the dissected land along the Wisconsin River. North of the Wisconsin River is the highly dissected hilly upland. (Data—National Oceanic and Atmospheric Administration; reproduction—EROS Data Center)

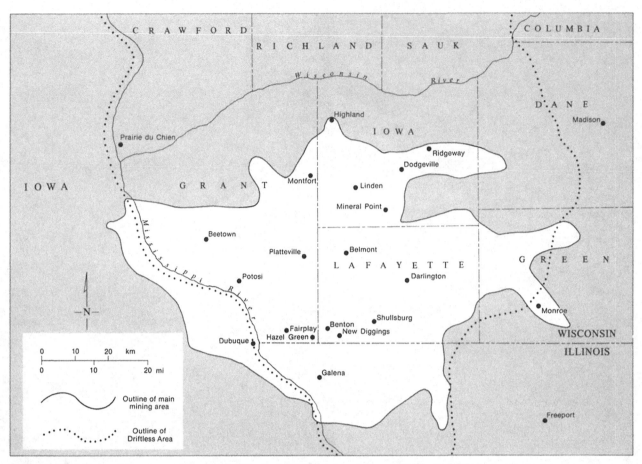

Figure 5.29 Main mining area of the Upper Mississippi Valley Lead-Zinc District. (U.S. Geological Survey)

The lead and zinc deposits for which this area is renowned are mainly in the Galena Dolomite and in the dolomites and limestones of the Decorah Formation and the upper part of the Platteville (fig. 4.31). Smaller deposits have been found in the Prairie du Chien Group. These are all of Ordovician age. Less significant amounts have been found in Cambrian and Silurian formations. Exploratory drilling is being done to locate more minable deposits which undoubtedly exist in extensions of the lead-and-zinc-bearing formations outside of this district, concealed by overlying rock or glacial drift.

The lead and zinc ore of southwestern Wisconsin was deposited by mineral-bearing solutions in scattered, irregular openings in the horizontal rock strata. These openings were created mainly by the dissolving and removing of the carbonate rock by groundwater. The ore impregnated the rock, which had been subject to only gentle bends and small faults. It filled crevices and empty spaces along joints, bedding planes, faults, and folds, and it lined cavelike apertures. As a result, the ore deposits were variously shaped—linear, arcuate, honeycombed, branching, vertical or horizontal, thick or thin. The largest were a mile or so long; but most were smaller and discontinuous, some being just tiny pockets and narrow seams.

The presence of the ore deposits brought to this area cultural influences that still exist, and their location helped determine early patterns of settlement and transportation.

Lead was sought before zinc was, and it was easier to obtain.

Lead mines used to be common sights throughout the southern part of the Driftless Area of Wisconsin. Some were north of the Lower Wisconsin River, but the majority were south of the river where the ore was more abundant because erosion of the ore-bearing strata has not progressed as far. The Wisconsin counties that led in production were Grant, Lafayette and Iowa.

One of the main lead-mining subdistricts encompassed Hazel Green, New Diggings, Benton, and Shullsburg in southern Grant and southwestern Lafayette counties, and extended northwest to the Platteville area. Another extended from southwestern Iowa County northeast to Ridgeway, including the Mineral Point, Dodgeville and Linden areas. Other important areas were around Potosi, Fair Play and Beetown in southwestern Grant County; and around Montfort and Highland on and near the border of Grant and Iowa counties (fig. 5.29).

Galena, the principal ore of lead, has been designated Wisconsin's official state mineral. It is lead sul-

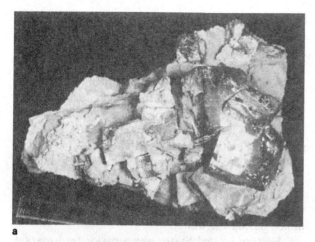

Figure 5.30 (a) Galena, the principal ore of lead, and Wisconsin's state mineral, has cubic cleavage, and when freshly exposed has a shiny metallic luster. The rock shown is 5 inches across. (Ralph V. Boyer) (b) A specimen that has dulled with time. The line scale = 5 cm (2 in.). (Geology Museum, University of Wisconsin-Madison. Photographer, Lawrence D. Lynch)

fide, containing 86.6 percent lead and 13.4 percent sulfur, and it is characterized by high specific gravity, perfect cubic cleavage, gray color, and, when freshly broken, a bright metallic luster (fig. 5.30).

Indians were the first "miners" in the district. They picked up pieces of "float" galena lying on the surface and also dug for it. They used it for decorative or ceremonial purposes. Cubes of it have been found at old village and burial sites. However, the amount used was small, and probably no smelting was done until the French arrived. The French, as individuals, did most of the first geologic explorations in the lead district in the 1700s. Lead joined furs as an exchange commodity. It was in great demand, especially for making shot for guns in this region abounding with wild game.

The earliest non-Indian newcomers to southwestern Wisconsin were able, like the Indians, to simply collect loose pieces of lead from the ground surface. The lead could be recognized by its heaviness, its luster if newly broken, and its cubic shape if not too smoothed

by abrasion. It had lain there after the rock which had enclosed it had weathered away, or it had fallen down slopes from exposed deposits uphill. Tumbled-down pieces were a clue to where desposits were exposed upslope, and such outcrops were quickly located. Because the land was well dissected, exposures were many. Linear depressions at the surface, indicating a crevice below, were also signs of possible ore deposits; lead was commonly found near the surface in crevices that tapered downward.

Settling of Wisconsin began in 1819. News of the widespread, easily obtained lead then brought thousands of Easterners and Southerners into southwestern Wisconsin, arriving mainly by way of the Mississippi River. The "Lead Rush" was on. The state's first permanent mining camps were established in 1824 at New Diggings and Hazel Green, then called Hardscrabble. About that time the fur trade was diminishing, and in 1832 the Black Hawk War marked the close of Indian warfare.

Mineral Point became an important mining center. It is on a point of land between two branches of the Pecatonica River where not only lead was found, but also copper, iron and zinc. In its early days Mineral Point was, for a time, the largest town in Wisconsin.

Lead production increased as people kept coming to that mineral-rich corner of the state. The population there grew rapidly from about 200 in 1825 to 4,000 in 1826, and 10,000 a few years later.

Mining was done at many sites in a quick, crude way, using a pick and shovel. A miner would simply dig a pit and reach for the ore in all directions as far as he could easily dig, then move to another promising spot. Often the ore was loosely embedded in easily worked clay or softened, weathered rock (fig. 5.31).

Figure 5.31 Holes from which lead was dug pitted the ground in many parts of the lead-mining district of southwestern Wisconsin following the Lead Rush, as in this unidentified location in Grant County. (State Historical Society of Wisconsin)

123

The first simple smelting furnaces were built of logs and stones on hillsides. The fuel was wood. Molten lead drained into a pit where it cooled in forms known as "plats." Thus the city of Platteville (formerly Platville) and the Platte River in Grant County got their names.

Energetic workers, with good luck, could make a small fortune during a summer of collecting and digging lead. It is said that a piece as large as a grapefruit was equal to a day's wages. Some miners would leave at summer's end but others stayed on. Over winter many of them found shelter in caves, or in pits from which they had mined lead, converting them into makeshift temporary dwellings or sod huts. These miners were nicknamed "badgers" because those animals are exceptionally good at burrowing into the ground and they hibernate during much of the winter. So Wisconsin came to be called the Badger State (fig. 5.32).

The hillsides of southwestern Wisconsin became pock-marked with mining pits that were dug and abandoned, and around which piles of waste were left. Many of the thousands of old pits and dumps have been smoothed over and erased by grading and plowing, or have become smaller as a result of natural filling. The largest are the size of a house basement. Only a few hundred pits are still identifiable. Trees and bushes generally grow over the remaining ones that cannot be cultivated or pastured.

Between 1830 and 1850 there was an influx of thousands of miners from Cornwall, England. Their coming was spurred by the depletion of Cornwall's mines and news that in Wisconsin "lead ore was sticking out of the ground," and it coincided with a new phase of mining in the Upper Mississippi Valley District.

The easily acquired galena had been taken—that gathered from the surface and dug from the shallow pits. Now to reach more deeply buried ore it was necessary to use explosives, build shafts, and work below the water table where flooding had to be controlled. The Cornish were experienced in those methods; so also were the Welsh and Irish who came. The heritage of those mining families is present today in the region's culture.

Skilled stonemasons built fine houses of dolomite and sandstone, many of which are still to be seen in Mineral Point and other places in the lead-mining regions (figs. 5.33, 5.34).

Something in the geologic makeup of the area drew the Swiss too—not so much the ore as the topography. They were agriculturalists, and the verdant hills around New Glarus in northwestern Green County bore a resemblance to their homeland. There they settled in 1845, and there their influence is vividly apparent today.

Agriculture was growing and diversifying throughout southwestern Wisconsin as the population increased. Some people farmed during the summer and mined during the winter.

Eastern United States was a major lead market, so much was sent there. Rivers were the main shipping routes, and almost every river port had its own smelter.

Figure 5.33 A typical cottage of a Cornish miner's family (Ingraham House) on Shake Rag Street in Mineral Point, built about 1840 of local sandstone. Stones for the front were cut more evenly and smoothly than those for the side. (Richard W. E. Perrin)

Figure 5.34 Brisbois House, a historic site at Prairie du Chien, is one of the oldest stone houses in Wisconsin, built in 1815. (Wisconsin Department of Natural Resources)

Figure 5.32 An early "badger" type lead mine near Highland, western Iowa County. (Wisconsin Geological and Natural History Survey)

The shallow Wisconsin River with its many sandbars could not accommodate large boats, but was navigable for smaller ones. River boats took lead down the Mississippi to New Orleans, or up the Ohio, for transshipment to the east coast. The lead trade helped to finance river steamboats of that time.

By the time the Wisconsin Territory was established in 1836, growing ports on Lake Michigan were competing for the lucrative lead-shipping trade. Large wagons called "lead schooners," drawn by eight to sixteen oxen, hauled pig lead from the mining district to lake ports. During the summer of 1841, 20 to 30 heavily weighted wagons arrived in Milwaukee every week, each wagon bringing about two and a half tons of lead. With the continued arrival of miners from Cornwall, lead production in the Upper Mississippi Valley District was soaring. In 1847, 13,000 wagons rumbled into Milwaukee with about 30,000 tons of lead to be shipped to Buffalo, New York and London. This was a larger lead trade than New Orleans had ever handled. As a result, road construction improved and a railroad was built from southwestern Wisconsin to Milwaukee.

One vestige of the early lead-mining era is the shot tower near Helena on the south bank of the Wisconsin River, preserved for its historic value in Tower Hill State Park. It operated for about thirty years, until the start of the Civil War in 1861 when lead was becoming harder to mine. There shot for guns, in various sizes, was made from finished lead. The shot tower was constructed by digging a 120-foot vertical shaft down through a riverside bluff. Shot was made by pouring molten lead through holes of a ladle, letting it fall down the shaft. The holes were of different sizes, and so were the lead droplets that fell. As they fell they formed into spheres, and hardened when they landed in a tank of cold water at the bottom. About one-sixth of the lead balls were good. They were put through buckskin sieves to sort them according to size. Imperfect balls were re-melted (figs. 5.35 to 5.39).

Wisconsin's lead production reached its high level in the 1840s, when it amounted to more than half of the nation's output. It peaked around 1845. A few years later the Gold Rush drew roving miners west; and agriculture, which had developed partly to support the mining production, began to be the main occupation in southwestern Wisconsin. Still, mining continued, and many miners who had left drifted back. Explorations reached farther underground for more rich deposits.

Wisconsin achieved statehood in 1848. On the right side of its official coat of arms appear a miner, a pick and shovel, and a pyramidal pile of plats. At its top is a badger. (See title page at front of book.)

Interest in mining zinc in southwestern Wisconsin began about 1859, as the market for it was growing. The zinc ores lay deeper than the lead ores, usually below the water table; and they required a more com-

Figure 5.35 The top of the old shot tower at Tower Hill State Park. (Wisconsin Department of Natural Resources)

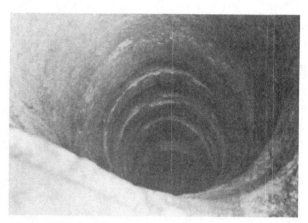

Figure 5.36 Looking down the hole of the shot tower, where molten lead was dropped. (Milwaukee Public Museum)

plicated refining process. They are of relatively low grade, but mining of them was helped by modernized techniques. After 1860 zinc production surpassed that of lead in this district. A zinc plant at Mineral Point, built in 1886, was for a time the largest in the world.

Zinc is a component of brass and other alloys, and is used for galvanizing and in making many industrial and chemical products. Lead too is used for alloys, and for such things as paints, ammunition, construction

Figure 5.37 Looking out of the tunnel at the foot of the shot tower. (Milwaukee Public Museum)

Figure 5.38 The bottom of the shot tower where cooled lead balls of shot were removed. (Wisconsin Department of Natural Resources. Photographer, Galen Parker)

Figure 5.39 From Tower Hill one looks across the Wisconsin River Valley to hills of the Western Upland on the other side. The river with its sandbars and islands was fine to travel on in canoes and small boats, but large boats had trouble navigating through the maze and shallow parts. (Wisconsin Department of Natural Resources)

materials, weights, and batteries. It has special uses because it is a soft metal, easy to work; is highly resistant to corrosion; is unusually heavy; and has both a low melting point and a high boiling point.

Lead and zinc brought industrial prominence to Wisconsin's southwestern upland for a while, but that prominence waned. The region contains no large industrial centers. Over the years mining operations became highly mechanized but production fluctuated markedly. The main periods of lead and zinc mining occurred when industrial demands were high, particularly during times of war. There were times, too, when production was just small-scale, as during economic depressions, but even then there was always at least one mine operating. Southwestern Wisconsin had earned the distinction of being the oldest continuous mining district in the United States. Then in 1979, the first total cessation of mining occurred in this district, when all mines were closed, until economic conditions again would make mining profitable.

The Galena-Platteville Cuesta Region of Eastern Wisconsin

The geologic base strongly affected the course of events in the Galena-Platteville Cuesta area of southwestern Wisconsin, as we have seen, but as that cuesta curves to the northeast it changes in character, and so does the record of human activity associated with it. In eastern Wisconsin the cuesta is drift-covered and less noticeable; permanent settlement came later; lead and zinc mining was absent; and industrial development took a different course.

Leaving southwestern Wisconsin and following the cuesta east into the glaciated region, we cross the Sugar, Yahara and Rock river valleys where the Galena-Platteville Cuesta is deeply dissected or worn through (in eastern Green, central and western Rock, central and eastern Dane, and western Jefferson counties). (See plates 1 and 4.) Separated tracts of dolomite acting as cover rock contribute to hilliness there. The bedrock controls topography more where drift is thin, and less where drift is thick and uneven.

East of this hilly break the cuesta is intact again, but as it turns north it is quite unlike its southwestern Wisconsin section where it forms an upland (including Military Ridge). Toward the north, to Green Bay, the cuesta's dip slope forms a lowland; along its eastern margin stands the stronger, higher Niagara dolomite.

The Galena-Platteville Dolomite forms a west-facing escarpment in eastern Wisconsin which overlooks the narrow strip of underlying weak St. Peter Sandstone and the narrow dip slope of the Prairie du Chien Cuesta. But this escarpment does not stand high, except in a few locations. In some places it appears as only a low ledge; and it is concealed along most of its length by drift that is from 50 to more than 100 feet (30 m) thick. The escarpment shows through noticeably near Seymour (Outagamie County) and Ripon (northwestern Fond du Lac County) (fig. 5.40). In glens between Ripon and the eastern end of Green Lake streams tumble over the escarpment in rapids and falls. The irregular escarpment line then passes inconspicuously northward through Winnebago County, past Lake Butte des Morts; continues northeast diagonally through Outagamie County and west of Pulaski; and crosses the state line in the vicinity of Walsh, northwest of Marinette.

In eastern Wisconsin, where the north-south cuestaform belts of alternating ridges and lowlands parallel one another, the Galena-Platteville Cuesta is sandwiched between the Prairie du Chien Cuesta on the west and the Niagara Cuesta on the east (fig. 5.5). Because the Prairie du Chien Cuesta's dip-slope lowland is narrow and blurred by the drift cover, it is generally disregarded in the bolder, linear ridge-and-lowland pattern; and the wider, more obvious Galena-Platteville Cuesta's dip slope is viewed as the second major lowland outward from the shield.

Widening of the Galena-Platteville lowland in eastern Wisconsin was helped by the weakness of the overlying Maquoketa Shale which was easily eroded downslope to the east, off the dolomite cuesta's dip

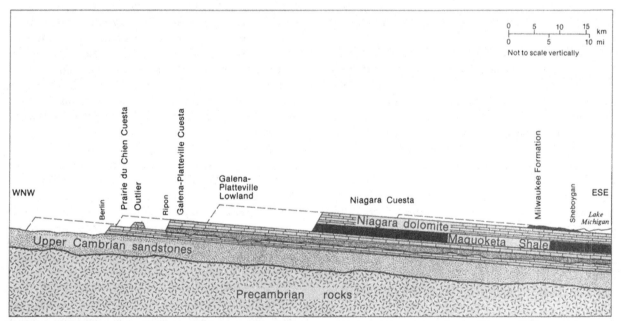

Figure 5.40 Diagram from Berlin to Sheboygan showing cuestas and rock formations in the Eastern Ridges and Lowlands. Dashed lines indicate former extensions of rock layers. Glacial drift is not shown. Not to scale vertically. (After Rex Peterson)

slope. Before glaciation the lowland had been formed by weathering and stream erosion; a river probably flowed along at least part of the lowland and through the Green Bay depression. Then the ice that slid from the northeast to the south eroded the soft rocks still more, smoothed down the escarpments in its path and removed most of the outliers.

Because the strata dip to the east here, it is along the lower, eastern margin of the Galena-Platteville lowland that erosive forces concentrated their work and created depressions, which drainage accentuates. Located there are such features as Green Bay, the Lower Fox River, Lake Winnebago, Horicon Marsh, and the Rock River.

The Galena-Platteville lowland is nearly 300 feet lower at Green Bay than at the Illinois border. It is about 20 miles wide, varying in width according to irregularities in its border, widening in the south to include the Rock River Valley. It is sometimes referred to as the Green Bay–Lake Winnebago Lowland, or the Rock River–Lake Winnebago–Green Bay Lowland, or is given some other name using a combination of those features. Naming of cities in the lowlands helps to show its location: from north to south, Marinette, Oconto, Green Bay, Appleton, Menasha, Neenah, Oshkosh, Fond du Lac, Waupun, Beaver Dam, Watertown, Oconomowoc, Jefferson, Whitewater, Janesville, and Beloit.

The waters of Green Bay cover much of the northern section of the lowland. Along the bay's western shore the dip slope is an exceptionally smooth plain, partly because it was formerly a glacial lake bed (plate 2). The lowland's middle section, from the southern end of Green Bay to Dodge County, is mostly rather level. The southern section, drained by the Rock River, is hillier because of bedrock dissection and uneven drift deposition.

Writing more than 100 years ago, T. C. Chamberlin described the Rock River Valley as an extension of Green Bay, "the two forming one great excavated trough." He presciently pointed out that the lowland's commercial importance was "very considerable," saying, "The sagacious proprietors of the Chicago & Northwestern Railway easily perceived this and located nearly 200 miles of their road in this valley, thus securing an easy grade along a line of important towns, supported by an exceedingly rich agricultural region, and possessing some of the finest water powers of the interior."[4]

Bedrock topography has had a strong influence upon the physical and cultural geography of the Galena-Platteville lowland and the rest of eastern Wisconsin. The main bedrock highs and lows have been important determiners of the large-scale drainage pattern. Glacial deposition lessened bedrock control of drainage, but did not eliminate it. The arrangement of cuesta ridges and lowlands, by directing the flow of major rivers, determined where the earliest arteries of travel and communication would be, and later influenced the routing of highways and railroads. Most of Wisconsin's large cities today are in the Eastern Ridges and Lowlands region, some being at harbors along the lakeshore.

The Fox and Rock rivers could have taken short routes to Lake Michigan if the Niagara Escarpment had not blocked their way. Instead the Fox detours north to Green Bay, and the Rock flows south and then to the Mississippi.

Between Lake Winnebago and Green Bay, the Fox River is known as the Lower Fox. It exits from Lake Winnebago at the northwest corner, and soon, at Appleton, it turns east down the dip slope to the low side of the valley. Because the Lower Fox cut through the drift to bedrock, its channel had many rapids and drops. There were eight sets of rapids between Neenah-Menasha and De Pere before the channel was improved for navigation.

The Lower Fox drops almost 170 feet (50 m) in about 37 miles (60 km) from Lake Winnebago to Green Bay. Most of this descent occurs in the 15 miles from Lake Winnebago to Kaukauna, and especially in the 9-mile stretch from Appleton to Kaukauna, with large drops of 38 feet at both Appleton and Little Chute, and a larger drop at Kaukauna. Because of its series of drops where power can be generated, the Lower Fox is an outstanding producer of waterpower. No other river in Wisconsin produces so much power in so short a distance. In no other valley of the state are there so many cities with so much industry so close together.

In 1915 it was written by Ray Hughes Whitbeck, professor of physiography and geography at the University of Wisconsin: "Among American rivers it is true that the Fox ranks only as a small river and its valley as a small valley, yet for more than 200 years this river and valley occupied one of the commanding positions of the Northwest; the Fox will always hold a place in history far out of proportion to its size."[5]

During the fur trading era and until the railroad era in the 1860s, the most used transportation route between Lake Michigan and the Mississippi River was that of the Fox River to the easy portage to the Wisconsin River at Portage. Despite the rapids in the Lower Fox, traveling on that river was preferable to carrying furs and exchange goods on one's back. Light Indian canoes, fit for shallow water and narrow channels, were

4. T. C. Chamberlin, *Geology of Wisconsin*, vol. 2, p. 103.

5. Ray Hughes Whitbeck, *The Geography of the Fox-Winnebago Valley*, Wisconsin Geological and Natural History Survey, bulletin no. 42 (Madison: The State of Wisconsin, 1915), p. 5.

lifted and carried around rapids. Then came the thirty-foot-long French canoes manned by up to ten men. In 1825 the Erie Canal in New York was finished, which let boats from the east coast enter the Great Lakes, and still-larger boats were used in Wisconsin. When steamboats became common, navigation on the Fox was improved. Locks were built in the Lower Fox, and in 1856 the canal at Portage was in operation, connecting the Great Lakes and the Mississippi River system.

The abundant cheap waterpower in the Lower Fox Valley very early attracted industry, including flour-milling and the manufacture of furniture, other wood products, and charcoal. Smelters at Appleton and De Pere handled iron ore from Dodge County and Upper Michigan, and various industries using iron were established. Later paper mills appeared, and the manufacture of paper and paper products expanded into a leading industry along the Lower Fox. The area continues to grow in importance as one of the state's foremost industrial regions.

The Galena-Platteville lowland as a whole is outstanding also in that it is one of the state's most productive agricultural regions. Its generally smooth terrain and the large amounts of dolomite in the soil help make it so. These factors similarly help to provide good to excellent agricultural land on much of the Niagara Cuesta, the next cuesta to the east and outward from the Precambrian shield.

The Niagara Cuesta

The Niagara Cuesta is the most continuous, prominent cuesta in the state. Its escarpment, which arcs through eastern Wisconsin for more than 230 miles

(370 km), overlooks the waters of Green Bay and much of the rest of the Galena-Platteville lowland. From the escarpment its dip slope slants down to the east at a rate of about 12 feet per mile, ultimately becoming submerged beneath Lake Michigan. The Lake Michigan shore and the adjacent lake bed constitute the outermost cuestaform lowland belt of Wisconsin. Weak Devonian shales were removed from this area by stream erosion and later by ice erosion. Lake Michigan's basin was probably a river valley in pre-glacial time (figs. 5.2, 5.41).

The northern part of the Niagara Cuesta in Wisconsin, Door Peninsula, was shaped by streams and scouring ice. Only because of the Niagara dolomite's hardness does the 75-mile-long peninsula exist at all, for softer rocks on either side were removed. The bedrock surface in the northern part of the peninsula is rough, and outcrops of dolomite through thin drift are numerous. The tapering cuesta provides beautiful scenery of cliffs, water-worn caves, and gracefully curved bays, as well as many harbors, all of which help to make this a popular resort, boating and recreation region (figs. 5.42, 5.43).

As a thin finger between large water bodies, Door Peninsula has gained a reputation for clean air, and its water-moderated climate permits the raising of frost-sensitive crops like cherries and apples, for which the peninsula is well known. It is favored with several state parks: Rock Island State Park on Rock Island off the tip of the peninsula; Newport State Park on the peninsula's northeast end; Whitefish Dunes State Park at the middle of the east shore; Peninsula State Park on the west shore near Fish Creek and Ephraim; and Potawatomi State Park along Sturgeon Bay.

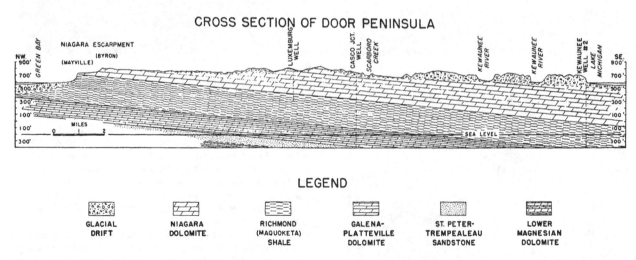

Figure 5.41 Cross-Section of Door Peninsula at its southern end. (Thwaites and Bertrand, *Pleistocene Geology of the Door Peninsula, Wisconsin*, Geological Society of America Bulletin, Vol. 68, 1957.)

Figure 5.42 From a panorama stop on Skyline Road in Peninsula State Park (on Door Peninsula west of Ephraim and north of Fish Creek) looking north, one sees the escarpment of the Niagara Cuesta following the indented shoreline along Green Bay with dark wooded cliffs and beaches of white dolomite cobblestones. (George Knudsen, Wisconsin Department of Natural Resources)

a

b

c

Figure 5.43 (a) Eagle Bluff in Peninsula State Park as seen from across Nicolet Bay. (Wisconsin Department of Natural Resources. Photographer, Staber Reese) (b) A closer view of Eagle Bluff, said to be the highest dolomite cliff in Wisconsin, shows the slowly eroding face of the jointed sedimentary strata. (Wisconsin Department of Natural Resources. Photographer, Dorothy Ferguson) (c) A cave in Eagle Bluff—like others along the shore, high above present water level—was wave-worn during glacial time when the Lake Michigan basin held a lake larger and deeper than today's lake. The lake level was about 30 feet (10 m) higher than today when this sea cave was eroded by waves. (Wisconsin Department of Natural Resources. Photographer, Staber Reese)

North of Sturgeon Bay the Niagara Escarpment, steepened by glacial erosion and wave cutting, has bluffs that rise 100 to 240 feet (75 m) above Green Bay. They would appear much taller if their base was visible instead of being covered by the water of the bay and the drift below it.

South of Sturgeon Bay the escarpment's west-facing bluffs are lower and farther inland from the shore of Green Bay than those to the north; and the peninsula's surface is undulating to gently rolling as drift becomes thicker to the south. State Highway 57 follows the cuesta's crest much of the way from Sturgeon Bay to the city of Green Bay. A narrow belt of Maquoketa Shale lies along the Green Bay shore of the southwestern part of the peninsula, at the base of the dolomite; it continues southward along the western edge of the cuesta (fig. 5.44).

Although the Niagara dolomite gives Door Peninsula beauty and strength, it also causes a problem which worsens as population increases—groundwater pollution. Throughout most of Wisconsin, groundwater is stored and filtered in the more porous and permeable rocks, or in drift or other gravelly, sandy surface material. But drift and loose earth materials are thin over most of the northern three-fourths of Door Peninsula, where dolomite is the main supplier of well water. That rock, like limestone, contains many fractures and wide openings into which polluted surface drainage may flow directly, and through which water moves without being naturally filtered and cleansed. This problem is compounded by the presence of a large population. In much of the peninsula, therefore, pollution of local groundwater is a constant concern to users, and engineering problems hamper the construction and operation of water and sewer systems, septic tanks and wells.

South of Door Peninsula, the Niagara Escarpment rises to its greatest prominence along the east side of Lake Winnebago, where it is known as "The Ledge" and forms what writers have called "mural cliffs." It descends sharply to the lake, dropping 223 feet (68 m) at High Cliff State Park, and 313 feet (95 m) south of Stockbridge. The old Military Road connecting the forts at Green Bay, Portage and Prairie du Chien was laid out along the crest of the Niagara Escarpment east of Lake Winnebago. A route west of the lake would have been shorter, but road construction was easier on the level ridge than on the less even, partly marshy lowland (figs. 5.45, 5.46, 5.47).

South of Lake Winnebago the escarpment edge is more irregular. It appears in outcrops east of Horicon Marsh and at Mayville and Iron Ridge. In Washington, Waukesha and Walworth counties it is covered by generally heavy drift, and outcrops are few.

As a result of glacial scour, the Niagara Escarpment in eastern Wisconsin, whether buried under drift or exposed, has a cleaner edge than the state's unglaciated escarpments, with fewer salients, indentations and outliers.

Chambers Island, 7 miles west of the escarpment in Green Bay, is an outlier. Other well-known Niagara dolomite outliers are in the Driftless Area capping mounds that stand on the Galena-Platteville Cuesta. They were left when the Niagara Cuesta retreated into Illinois and Iowa.

Blue Mound is the most famous of those mounds that are outliers of the Niagara Cuesta. It is considered by some to be the most conspicuous individual feature in the Wisconsin landscape (fig. 5.48). It is the highest point in southern Wisconsin, rising 1,719 feet (524 m) above sea level, or 415 feet (125 m) above the general land surface. Looking like a broad, truncated cone, it

Figure 5.44 The Niagara Escarpment, back from the shore in southern Door County. It is used here as a ski hill. (Ralph V. Boyer)

Figure 5.45 The northern part of High Cliff, including the state park, overlooking the northeast shore of Lake Winnebago. High Cliff is a section of the Niagara Escarpment, which is known as "The Ledge" in this area. (Wisconsin Department of Natural Resources. Photographer, Dean Tvedt)

Figure 5.46 In a snowy landscape the higher and more exposed sections of the Niagara Escarpment appear as dark arcs trending north, from the east (right) side of ice-covered Lake De Neveu in the central foreground, on along Lake Winnebago which is at the upper left. (Carl Guell)

Figure 5.47 The Niagara Escarpment is left wooded while the prime agricultural land on either side is cultivated. Here one looks north toward Oakfield, which is northeast of Waupun. County Highway B runs along the left (west) side of the escarpment. (Carl Guell)

is easily recognized from twenty or more miles away, and is visible from north of the Wisconsin River. So important was this landmark to early travelers that even as far away as Milwaukee a main street leading west is named Blue Mound Road. Today this distinctively shaped hill serves as an important reference point for pilots of small aircraft. The flat top of Blue Mound is composed of cherty rock more than 100 feet thick. Most of the Silurian dolomite in that cap rock was replaced by quartz, making the summit exceptionally resistant and thus delaying erosion of the mound. Weak Maquoketa Shale creates the slope immediately below the resistant cap, and older Sinnipee Dolomite forms the broader base below that.

Alongside Blue Mound stands smaller East Blue Mound, whose elevation is 1,489 feet (454 m) above sea level. Together they are sometimes referred to as Blue Mounds. The Indians called these mounds Smoky Mountains, probably because they look blue-gray in a haze, and are high enough to be enveloped in clouds at times. East Blue Mound has already lost its dolomite

cap and is therefore wasting faster. These mounds are 20 miles (32 km) west of Madison, just north of Highway 18 (the old Military Road) on the border of Dane and Iowa counties. East Blue Mound, in Dane County, is the site of a county park; and larger West Blue Mound, in Iowa County, of a state park. Both mounds contain caves.

The largest of the mounds northeast of Platteville, in northwestern Lafayette County, are also outliers of the Niagara Cuesta, retaining their dolomite caps. Double-peaked Platte Mound (or Platte Mounds), which is visible from far away, lies about 4 miles northeast of Platteville. Its highest elevation is 1,430 feet (436 m) above sea level, about 230 feet (70 m) above its surroundings. On its south slope is a huge letter M (for "mining") made of 400 tons of dolomite rock, which was constructed years ago by students of mining at Platteville's old Institute of Technology, later incorporated in the University of Wisconsin–Platteville (fig. 5.49).

Figure 5.48 Blue Mound, the highest elevation in southern Wisconsin and a landmark recognizable by its flat top and long sloping sides. (Wisconsin Department of Natural Resources)

Figure 5.49 Platte Mound bears the letter M (standing for mining), built with dolomite rocks by students of mining technology at University of Wisconsin-Platteville. (Carl Guell)

Two and a half miles (4 km) to the east is conical Belmont Mound, about 200 feet high. Between Platte and Belmont mounds is a third mound which has lost its dolomite cap and is much smaller than the others. Together the three are sometimes referred to as the Platte mounds. Wisconsin Territory's first capital was located between the two larger mounds near Leslie, at what was then the town of Belmont. That site is now a state park. Much of the early town of Belmont later moved south about three miles when a railroad line was built there.

Other well-known Niagara outliers in southwestern Wisconsin are Sinsinawa Mound in southern Grant County, which rises 185 feet (56 m) above its surroundings, and White Oak Mound in southern Lafayette County, about 200 feet high.

The Niagara Escarpment in eastern Wisconsin has no visible gaps except at the narrowing end of Door Peninsula, where a channel separates Rock Island from Washington Island, and Porte des Morts (Death's Door) separates Washington Island from the peninsula. However, there are deep cuts in the bedrock at Ellison Bay and at Sturgeon Bay. Those transverse gaps were probably made by streams before glaciation. Perhaps the Menominee River used to flow through Sturgeon Bay. That bay's eastern end is now connected to Lake Michigan by a canal one and a quarter miles long.

South of Sturgeon Bay the Niagara Cuesta is unbroken except for a drift-filled valley northeast of Lake Winnebago, which a tributary of the Manitowoc River now follows. All main streams of the cuesta's dip slope flow to Lake Michigan. No large stream flows in either direction across the escarpment. As a result, natural waterpower sites are few and small-scale in most of the Niagara Cuesta region of eastern Wisconsin; early industries there were dependent upon coal for power. Since coal was imported by way of Lake Michigan, industries became established at or near the harbors on the lake shore, mainly in the Milwaukee area and at Kenosha, Racine, Sheboygan, and Manitowoc.

Niagara dolomite is much desired for construction and ornamental purposes. This pale, yellowish-white rock is attractive, extremely hard, and long-lasting. In eastern Wisconsin it has been quarried almost everywhere it showed through the drift cover, which was mainly at the escarpment and along streams; and in many places where drift cover was thin, that overburden was scraped off to make the dolomite accessible (fig. 5.50).

"Lannon stone" from northeastern Waukesha County is Wisconsin's leading dolomite building stone. Certain other high-quality Niagara dolomites, like those of Fond du Lac and Calumet counties, are also in demand because of their color and hardness. Although other Wisconsin dolomites are also used as building stone, they generally are less hard, less durable, and less uniform in texture than the high-grade Niagara dolomite.

Other Resource Values of the Paleozoic Rocks

Because dolomite is widespread throughout eastern and southern Wisconsin and the Western Upland, it has been one of the state's main geologic resources. All of the state's dolomite formations—Devonian and older—have been much quarried. Durable dolomites that have a uniform texture can be trimmed, or shaped, well; so they are usually cut for building stone, while those dolomites and limestones that are softer or less homogeneous in texture are crushed for road and construction material, and for concrete aggregate. Crushed dolomite is added to iron ore as a flux in smelting, and is used as a furnace liner in steel manufacturing.

One of the earliest, most common uses of dolomite and limestone was to make lime, which was needed in quantity for mortar and plaster when this region was being settled. Many families made their own lime by burning the rock in small kilns to drive off the carbon dioxide, water and other volatile materials. Some kilns were quite crude, no more than a dome of boulders. There were large commercial kilns too (fig. 5.51). Natural cement made from the lime was used to mortar logs, field stone, cut stone, and bricks in general construction work. When portland cement largely replaced this natural cement, most kilns were abandoned, since dolomite cannot be used in making portland cement because of its excessive magnesium content. Lime continues to be indispensible in paper-making, tanning, smelting, and food-processing; in water purification and other sanitation processes; in the

a

b

Figure 5.50 (a) This church at Kirchhayn, eastern Washington County, was built in 1856 of Niagara dolomite quarried nearby. (Richard W. E. Perrin) **(b)** The stone beneath the windows contains brachiopod fossils. The mortar between the stones was made from local rock. (Richard W. E. Perrin)

manufacture of steel, paint, plastics, and other chemical and industrial products; and in agriculture as a plant food and a neutralizer for acid soils.

Wisconsin's sandstones also have a variety of uses. As mentioned before, the better-cemented sandstones, especially those with attractive coloration, are cut for building stone; while those that are weakly cemented can be crushed into loose sand for many purposes.

Figure 5.51 The Bauer lime kiln buildings near Knowles (northeastern Dodge County) were built about 1900 with local dolomite, and are seen here in a state of partial ruin. The stone crusher is the tall building at the right, and the kilns where the rock was burned are at the left. (Richard W. E. Perrin)

Wisconsin has vast reserves of homogeneous, high-purity quartz sand in its Cambrian and Ordovician sandstones, which yield grains in a broad range of desired sizes. Because the grains are hard and have high heat resistance, they can be used in foundries to form molds for casting metals. The availability of these sands has helped to make southeastern Wisconsin a leading center for metal foundry work and hence of machinery manufacturing. Sands are also needed for abrasives, sand-blasting work, filtration, and construction, and have various other industrial uses. Sands that are especially pure can be used to manufacture glass.

Shale, particularly Maquoketa Shale, provides clay for the manufacture of brick and tile, including drainage pipes.

There are indications that oil and gas may exist in some Paleozoic formations in Wisconsin, including the younger strata under the Wisconsin section of Lake Michigan, but the most promising locations and the quantity are uncertain.

Small quantities of iron ore are found locally in Wisconsin's Paleozoic sedimentary rocks. In eastern Dodge County mining of iron ore in the Neda Formation began in 1849. It was mined mainly at Iron Ridge, where the formation was about 20 feet thick. The ore was in beds several inches thick, conveniently exposed in a ledge that ran north-south, facing west. Neda ore was mined also at the Mayville Mine about three-fourths of a mile to the north. Because the ore was oolitic, or granular, it was easily dug and reduced, but the quantity available was not large. Smelting was done locally; it was economically feasible because dolomite, needed in the smelting process, was in the immediate vicinity. Mining was discontinued in 1928 after the largest accessible deposits were used, because mining the ore at depth was expensive, deposits were scattered and small, and the phosphate content was so high as to interfere with the processing of the ore.

Near Ironton, northwest of the Baraboo Hills in Sauk County, enough iron ore was mined to supply a small blast furnace at Ironton from 1850 to 1880. It produced about 11,000 tons of iron. The ore was hematite, which probably formed from iron that had been released as the rock holding it decomposed, and had been transported by solution into fractures in the Cambrian sandstone, where it was deposited. Similar iron deposits in Richland, Vernon, Crawford, and Pierce counties appear to have no economic importance.

Among the most important functions of Wisconsin's Paleozoic rocks are those of storing, transmitting, and purifying groundwater. Precipitation is absorbed by the rocks, and eventually the water is discharged naturally into rivers and lakes through seepage and springs, or is withdrawn from wells. One reason Wisconsin is generally favored with a copious, reliable and high-quality natural water supply is that its sedimentary rocks perform this function well.

All of Wisconsin's main sedimentary formations, except Maquoketa Shale, are aquifers—that is, "water-bearers," or rock units that transmit groundwater. Some are better aquifers than others. The Ordovician dolomites contain large supplies of water except where they are deeply buried and where impermeable Maquoketa Shale is present above. Sandstones, whose interconnected pore space allows freedom of water movement, are the best aquifers in terms of volume held, permeability, and purifying capability.

In Wisconsin, Cambrian sandstones contribute the most water to springs and wells, and are the most heavily drawn-upon aquifers. They supply large, dependable quantities of water, which is generally of desirable quality, although water from deep sources may be highly mineralized. These sandstones are present in significant thickness in over half of the state, including its most densely populated and heavily industrialized areas. They fill the public water requirements for many of Wisconsin's cities, supplement surface-water supplies in others, and supply numerous rural wells. The most productive sandstones are the Mount Simon and Galesville. The amount of water contributed by the St. Peter depends mainly upon its thickness, which varies greatly from place to place.

Certain springs and wells in Wisconsin became known long ago for the special mineral content of their water, which was believed to have therapeutic properties when consumed and bathed in. People would come to them to obtain water for their personal use. The Waukesha area has been especially renowned for its mineral waters, and enjoyed resort status in years past because of them. In the early 1900s spring and

Figure 5.52 Old bottles once used by companies selling natural water, like these from Wisconsin, are now collectors' items. (Roger Peters)

Figure 5.53 Rock in the process of being removed by groundwater solution in Bear Creek Cave, near Plain in southwestern Sauk County. Some of the dissolved limy material is precipitated by the groundwater to form stony "waterfalls" called flowstone. Thin "soda straw" stalactites are left by water dripping from the ceiling. Access to this cave is through a quarry. (Richard J. Boyd)

well water was being sold by companies in many parts of the state, mainly in the Paleozoic regions—in or near the cities of Oconto, Green Bay, Chippewa Falls, Osceola, Sheboygan, Maribel, Prairie du Chien, Darlington, Madison, Waupaca, Oshkosh, Green Lake, Wauwatosa, Menomonee Falls, and at places in Walworth, Racine and Waukesha counties, as well as other locations. At many of these places the waters were bottled to be sold, or used in the making of soda water and other beverages. At one time Wisconsin led all states in water sales. Its water was shipped as far away as the east coast and Florida, and was sent by tanker loads to Illinois (fig. 5.52). The sale of mineral water for beverage-making and personal consumption declined as a business, but recently the market for pure natural water has been increasing.

Caves

Dolomite bedrock, like limestone, characteristically contains caves, and Wisconsin has its share. As groundwater moves through the rock, dissolving and removing lime, it slowly enlarges cracks to crawl-space and tunnel size, and some eventually become the size of a room or several rooms. Of the few hundred caves known in Wisconsin, none is larger than that or more than 2,000 feet (600 m) long. Most are quite small, but they are of interest and value (fig. 5.53). In regions outside the state where limestone and dolomite formations are much thicker and where cave-forming processes have been at work longer, caves are larger and more numerous.

Many cave entrances are sinkholes—dissolved openings or cave-ins at the surface through which water drains down into subterranean passages. Most sink-

holes in Wisconsin are not more than 15 feet (5 m) in diameter, but some are several times as large. Water may drain through them without obstruction, or, if drainage is blocked, it may form a pond in the hollow and drain slowly.

Cave entrances also may be on hillsides or cliff faces. Certainly numerous caves are unknown because they do not have an opening to the outside. Some that are said to have existed, according to old writings, memory or legend, can no longer be located because their entrances have collapsed or been covered. Quarrying sometimes reveals unsuspected caves.

Caves provide sport for explorers and an opportunity for geologic study underground. Several of the large Wisconsin caves are open to visitors, allowing them to see the artistic formations created by groundwater. Some caves are closed to visitors to protect their fragile natural beauty. Long ago sunlit cave openings

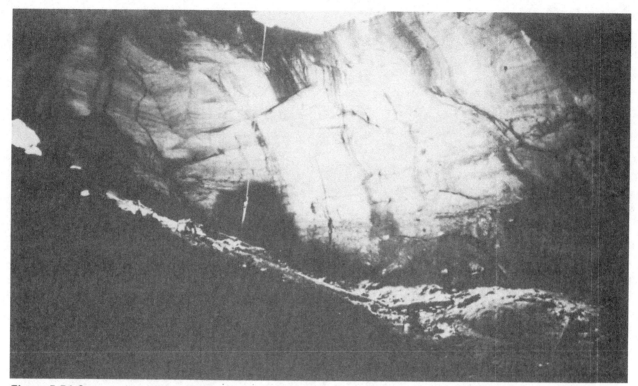

Figure 5.54 Cave explorers enter Bridgeport Cave through a sinkhole at the top and rappel down a rope in a 40-foot drop to the slanty, sandy floor. The walls of this cave are sandstone. Dolomite was above and below. The cave is northeast of Bridgeport in southern Crawford County, north of the Wisconsin River. (Richard J. Boyd)

Figure 5.55 The caves in Maribel Caves Scientific Area are small hollows in a dolomite cliff. Animals sometimes use them as shelters. (Ralph V. Boyer)

(not the dark, damp recesses) served as shelters for nomadic people, especially in winter. Archeological evidence shows that some caves in the Driftless Area were occupied at least six or seven thousand years ago and at different times since then.

Most of Wisconsin's caves are in Prairie du Chien Dolomite, but other dolomites contain caves too. Some caves are found in sandstone strata, but the larger of those that are underground result not from removal of the sandstone but of the dolomite underlying it: a cave forms in the dolomite; its roof collapses; the cave fills with unsupported sandstone fallen from above; and the resulting hollow is above the dolomite formation, in the sandstone level (fig. 5.54).

In Paleozoic eastern Wisconsin, glaciers undoubtedly destroyed or removed most of the preglacial caves either by crushing them under their weight or by scraping off surface rock. Drift undoubtedly conceals the openings of most remaining caves there. Only a small number of caves have been found in the glaciated part of Wisconsin, several of which are in Door County. In Maribel Caves County Park in northern Manitowoc County there are small openings in the Niagara dolomite. They and the clear mineral springs of the Maribel Bluffs have long attracted visitors (fig. 5.55).

* * * * *

We have seen what significance the Paleozoic rocks have had in the modeling of Wisconsin's landscape and in the region's economic and cultural development.

In the Driftless Area erosion of those rocks went on without interruption. But the rest of Wisconsin was to experience yet another metamorphosis. It would figure prominently in that most recent of the world's great geologic-climatic events—Pleistocene glaciation.

6

Wisconsin's Ice Age Heritage

Nowhere in the world is the geologic story of the Ice Age more dramatically displayed than in the Wisconsin area. And one can hardly think of a region that has been more enhanced by glaciation if one considers not just scenery but the physical resource base as well.

In some other glaciated areas the landscape is visually enriched by mountains carved sharp and jagged by glaciers—the result of ice erosion. In contrast, Wisconsin's glacier-created assets result mainly from what the ice deposited. Though its glacial landform features are not as lofty and scenically breathtaking as ice-gouged, barren mountains, they are likewise appealing; and to many people they are of greater interest because of the clues they hold to riddles of the past. From the deposits left by a series of glaciers much can be learned about the movement and work of those vanished masses of ice (plate 2).

Indeed, Wisconsin is glacially world-famous. Here one can see, side by side, heavily glaciated country and the unglaciated Driftless Area, and thus observe how the ice changed the land. Understandably, people have come from far and wide to study glaciation in Wisconsin. The Ice Age National Scientific Reserve, the first reserve of its kind, was established in Wisconsin in 1973 to preserve and explain outstanding glacial features. The reserve consists of a number of separate parks and sites throughout the state, some of which will be discussed later (fig. 6.1).

Characteristics of Glaciers

Before looking more closely at glacial events and features in Wisconsin, it is well to review some basic facts about glaciers.

A *glacier* may be defined as a mass of ice that has formed naturally from compacted snow, and that moves or has moved by slow flowage. Gravity, of course, draws glaciers downslope, as it does drainage in liquid form. But glaciers have the ability to spread under the pressure of their own weight. So although glaciers gravitate

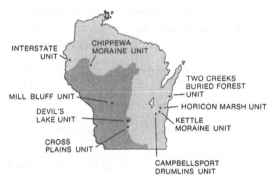

Figure 6.1 The units of the Ice Age National Scientific Reserve, which is operated by the National Park Service and the Wisconsin Department of Natural Resources. The lighter area of the map was glaciated in the last major ice advance.

to lowlands, as water does, they can, if thick enough, also "flood" across flat terrain, and can even move upslope when the pressure behind them is sufficient. That is why glaciers, unlike rivers, can erode material from well below their local base level, and even from below sea level as they have done in lakes Superior and Michigan, carry it to higher locations beyond, and there deposit it.

Glaciers keep moving and expanding as long as the amount of snow accumulating on them is enough to maintain the needed ice thickness and pressure. Then they can envelop or override all obstacles in their paths. But when melting occurs faster than snow accumulation, glaciers weaken. They become thinner and waste along their warming edges, though ice may still be flowing to those edges. If melting continues, the ice front "retreats."

When it is said that a glacier retreats, or withdraws, it is not meant, of course, that the glacier itself flows backward, but that its outer edge melts back faster than the body of ice moves forward. In its dying phase, however, the whole ice body may become stationary, and thin more and more, and finally be reduced to separated patches, shrinking away as a snowcover does at winter's end.

There are small glaciers of limited size, such as those confined to cold mountain valleys. There are complexes of merged glaciers covering broader areas, which are called *ice caps*. The largest of all glaciers are the *ice sheets*. Those that cover, or covered, sizable areas of continents are called *continental ice sheets*. It was such an ice sheet that glaciated Wisconsin.

That ice sheet, which is named the Laurentide Ice Sheet, originated in eastern Canada and spread outward in all directions, as explained in the chapter 1 synopsis. Like Pleistocene glaciers elsewhere, it came and it withdrew a number of times during the Ice Age.

The General Time Frame

A cold, snowy period when an ice sheet formed and spread is known as a *glacial stage*.[1] Stages were separated from one another by a warm *interglacial stage* during which the ice sheet temporarily melted away. It may have disappeared completely, or shrunk and retreated so far that its climatic effect was weakened. During interglacial stages climatic conditions in the glaciated regions, including Wisconsin, became much like those of today, with weather as warm or even warmer than it is now. Plants and animals reentered the area as the ice left. However, from one interglacial stage to the next, biotic communities were altered by extinctions, new adaptations, and new arrivals.

In the past, estimates of the duration of glacial stages within the Ice Age have ranged from tens of thousands of years to more than one hundred thousand years. It seems that stages lasted for different lengths of time, as did interglacial stages. Much remains to be learned about the duration of those cold and warm periods, especially the earlier ones (figs. 6.2 and 1.9).

1. Although specialists are recommending new alternate terms for "stage," none has become standard in usage. Longer-established, better-known terminology is adhered to here. For the general reader, the specialists' term "Wisconsinan" is likewise not used.

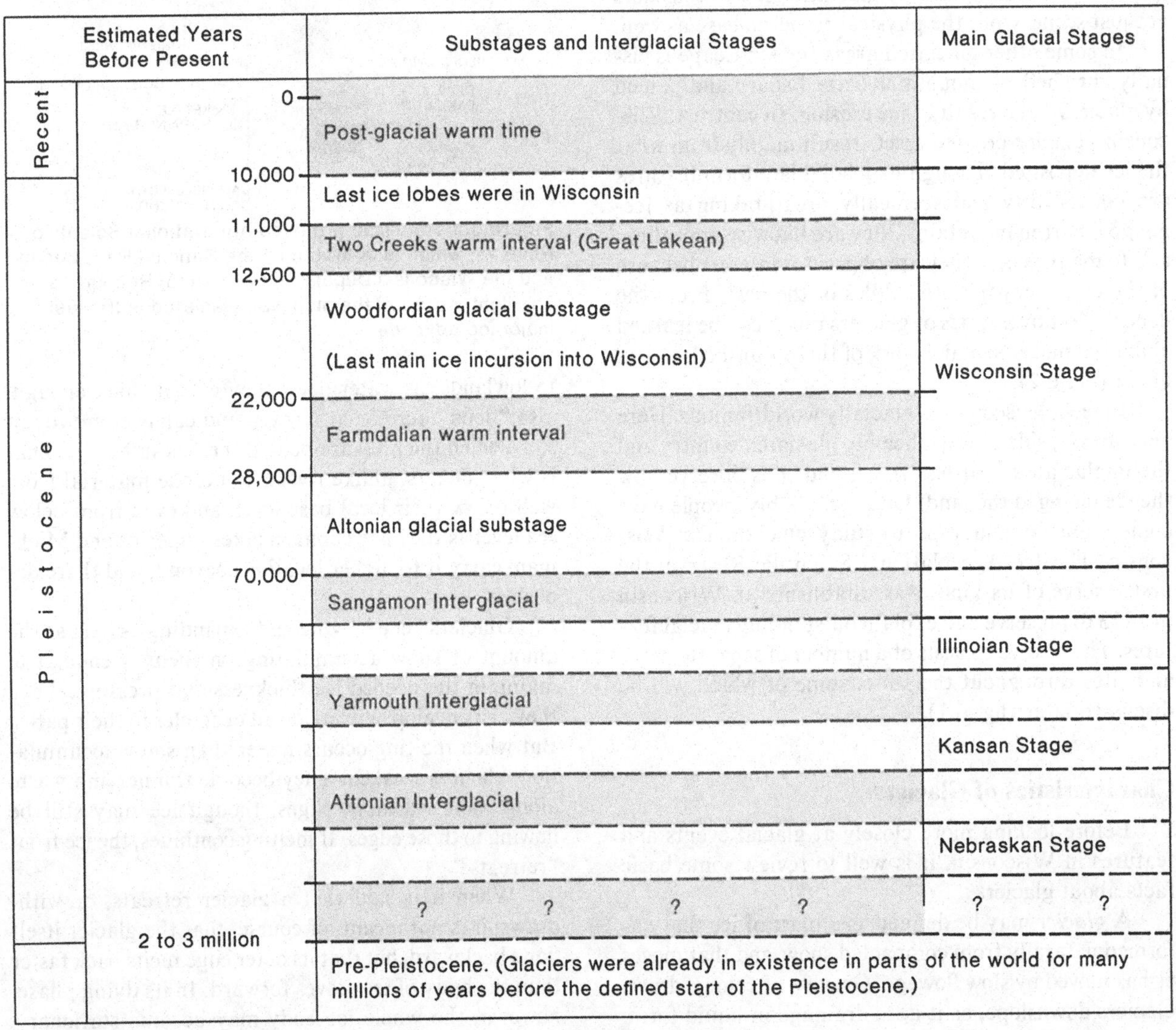

	Estimated Years Before Present	Substages and Interglacial Stages	Main Glacial Stages
Recent	0	Post-glacial warm time	
	10,000	Last ice lobes were in Wisconsin	
	11,000	Two Creeks warm interval (Great Lakean)	
	12,500	Woodfordian glacial substage (Last main ice incursion into Wisconsin)	Wisconsin Stage
Pleistocene	22,000	Farmdalian warm interval	
	28,000	Altonian glacial substage	
	70,000	Sangamon Interglacial	
			Illinoian Stage
		Yarmouth Interglacial	
			Kansan Stage
		Aftonian Interglacial	
			Nebraskan Stage
		? ? ? ?	? ?
	2 to 3 million	Pre-Pleistocene. (Glaciers were already in existence in parts of the world for many millions of years before the defined start of the Pleistocene.)	

Although the old, classic four-stage time scale still serves as a basis for study, it is now recognized that there were more glacial advances than the four perceived by early investigators.

Figure 6.2 A time scale of the Pleistocene Epoch in northcentral United States.

The warm Recent, or Holocene, time in which we live may be an interglacial stage of the Ice Age, or the Ice Age may have ended. Only time will tell. Ecological changes continue to occur while we move either closer to, or farther from, glacial conditions. We cannot say which; the period of recorded observation has been too short. Much theorizing is being done, but the fact remains: we cannot foretell future climate.

If we knew what caused the onset of worldwide glaciation at the start of the Ice Age, we would be better able to account for the waxing and waning of the ice sheets. The cause of the Ice Age remains unknown despite the popularization of one theory after another.

We do know that Wisconsin was invaded by the Laurentide Ice Sheet several times during the Ice Age (fig. 1.25). Each time the ice sheet took form in Canada and expanded, it must have gone through a series of preliminary, progressively stronger advances separated by setbacks. In other words, it must have spread not steadily but with a number of alternating advances and retreats as it grew ever larger, reaching toward its farthest position.

Times of pronounced individual advances or retreats within a stage are termed *substages*. Glacial deposits, including the arrangement of certain glacier-created landforms, show that ice sheets did go through a fluctuation of advances and retreats during their waning period. However, evidences of the preliminary thrusts made during their growing period were covered or erased as the ice kept coming and passing over them.

As a result of scarcity of data, and utter lack of data for long stretches of time, we cannot plot a line graph accurately showing the climatic variations of the Ice Age over the millenia (though helpful attempts have been made). If we could plot a temperature graph of the Ice Age, the line would not be smoothly oscillating with well-defined cold glacial dips and warm interglacial humps. Rather, because of the many substages and lesser fluctuations, it would show numerous irregular ups and downs; so in places it might not be clear whether a wavering section of the line should be considered part of a warming trend or a cooling trend, and sometimes it might be difficult to decide whether a fluctuation should be classified as major or minor—for example, whether a certain cold period was of the proportion of a short, weak glacial stage or a strong substage. Therefore, locating the divisions between glacial and interglacial stages, and assigning relative lengths of time to each, might not be a simple, clear-cut matter even if one did have the full climatic record.

Differentiating the Drift Layers

Translating this problem to the out-of-doors, we find that field workers encounter many complications as they analyze glacial deposits and other clues in at-tempts to reconstruct the timetable of the Pleistocene. Though it is the geologic time period closest to our own, it is, like more remote periods, rife with perplexities for the investigator.

Signs of individual stages and substages can be found in layers of drift left by successive ice sheets. When an ice sheet (or any glacier) melts away it leaves typical, identifying deposits on the land it covered: transported material that ranges in size from large boulders down to smaller gravel, sand, and clay particles. And these deposits are distributed in characteristic ways to be described later (figs. 6.3, 6.4).

If, in a given area, one distinguishes drift layers of, for example, three different ages, one knows that at least three separate times a glacier has been there. Drift deposits of different ages, one atop the other, may be exposed at the surface in small or large areas, or may be found by digging and boring into the subsurface. It is possible that some drift layers that were once present are no longer there or are unrecognizable, having been carried away, or mixed with new deposits when a later ice sheet overrode them. Drift layers that are near or next to each other may be differentiated (but not always easily) by several means, including their position, composition and apparent age.

The degree of weathering denotes relative age: the longer material has lain the longer it was weathered, and the effect of additional thousands of years is evident. Also, younger drift typically has a "fresher," or rougher, topography than older drift.

If a horizon, or layer, of buried soil exists within drift, it marks a separation between two layers of drift—an older lower layer and a younger overlying one. The ice deposited the lower layer, and then withdrew long enough for vegetation to grow on it and for soil containing vegetation's dark remains to develop on it. Then the climate cooled again, and the ice returned to deposit a new layer of drift over the soil.

Drift layers may differ noticeably in color—brown, reddish brown, gray. Various layers may contain unlike, distinguishing ingredients. They may have dissimilar mixes of earth materials, showing that the ice that dropped them had crossed different territory and bedrock. Where drift layers are similar in color and composition, telling them apart is difficult. In most cases, means are not available for conclusively comparing or matching ages of drift deposits if they are not adjacent or continuous, and often even if they are. Some maps differentiate drift deposits of unlike ages as only "new drift" and "old drift" for simplification, or because of uncertainty about the age of one or more layers.

The age of plant or animal materials buried in the drift—like bones, ashes, or parts of trees and plants—can sometimes be estimated by radiocarbon dating (if they have not been contaminated); and if these materials have not been moved since being deposited they

a

b

Figure 6.3 (a) Man observing unsorted, bouldery glacial drift. Langlade County. (Wisconsin Geological and Natural History Survey) **(b)** Glacial drift, unsorted. (Milwaukee Public Museum) **(c)** Water-sorted drift—fine materials deposited in layers. Near Muscoda. (Gene Musolf)

c

Figure 6.4 Different layers of glacial drift can be seen along the eroding shore of Lake Michigan at Cudahy. Gullies are cutting into the bluffs. Some of the darker drift is washing down over lighter drift. (Wisconsin Coastal Management Program and Wisconsin Department of Natural Resources)

142

can shed light on the age of the drift containing them. This dating method measures the radioactivity of a test specimen, and since radioactivity diminishes at a known rate after a plant or animal dies, the time when it died can be estimated. This test can be effectively used only for material younger than about 50,000 radiocarbon years; that is, from the last part of the Pleistocene. Wisconsin's drift has provided a number of radiocarbon dates from scattered sites.

Some of the first glacial stratigraphy studies were done in the European Alps, where researchers in the late 1880s distinguished deposits of four separate stages of local mountain glaciation. That four-stage pattern was then applied to the larger-scale glaciation of northern Europe which had been overrun repeatedly by the Scandinavian Ice Sheet. And when field research was done in central United States along the outer limits of the Laurentide Ice Sheet, where sections of overlapping drift layers are exposed, four main stages were designated there too. (Perhaps the classifying of the drift there into four stages was influenced by the knowledge of the previously designated four stages in the Alps.) The stages were named for states in which deposits of those stages were well displayed and where early field work on them had been done. From oldest to youngest they were called the Nebraskan, Kansan, Illinoian, and Wisconsin stages (fig. 6.2).

After early mappers had divided the North American ice-sheet deposits into four stages, other field researchers, with new techniques and fresh interpretations, formed different opinions and chose to reassign some exposures from one stage to another, and from one substage to another. In Wisconsin, as elsewhere, there has been considerable renaming and redating of glacial stratigraphic units. Differences of opinion exist, and the classification of many exposures, large and small, remains controversial or undetermined. So when one consults even the latest or most used classifications, chronologies, and maps of glacial geology, one should bear in mind that they are not universally accepted and will be revised as new findings and techniques become available.

Every time the Laurentide Ice Sheet expanded it assumed a somewhat different outline, though always it seems to have been spreading toward the same general limits. It is impossible to map the shape and extent of each of its main advances throughout the Pleistocene. The only stage for which its full dimensions are fairly clear is the last, the Wisconsin Stage, because its deposits are uppermost, on the surface.

The layering of sequential glacial deposits is comparable to a skirt with a number of uneven petticoats beneath it (fig. 6.5). Deposits of the Wisconsin Stage might be said to be the exterior, completely visible skirt on top. Its uneven hem swoops from the Canadian Rockies through Montana and the Dakotas, through Iowa and south of the Great Lakes, across Pennsylvania and New Jersey; and the Driftless Area is a wide slit or tear in that skirt. Drift layers deposited during earlier stages are ragged petticoats that droop out here and there around the skirt's hem. Earlier deposits show also in holes worn through one or more of the outer garments (holes eroded, dug or drilled through the overlying drift). Just by looking at such an uneven hemline (even with the help of the holes) we cannot count for certain how many full petticoats there really are beneath the skirt, nor tell precisely where in the sequence a particular sagging petticoat belongs. Some petticoats are everywhere shorter than the skirt, as a good petticoat should be, and do not show at all.

It is unlikely that the ice sheet, during any given period of expansion, covered all of its glaciated area at the same time. It changed shape continually. Surges of ice could come from different centers at different times—for example, from west of Hudson Bay, from east of it, or from other centers of growth. Certain segments of the ice sheet were at times climatically favored, growing faster than others. Some sectors could be wasting while others were expanding.

The Spreading of the Ice

The Laurentide Ice Sheet was larger than Europe's Scandinavian Ice Sheet, and overspread an area nearly as great as Antarctica's ice sheet does now.[2] It may have been as thick as Antarctica's ice sheet in its thickest parts. How thick it was over Wisconsin, near its margin, is not known. It did cover the bedrock hills, but the absence of high mountainous features in the state—in the whole center of the continent, in fact—leaves us without a tangible reference point for determining the height of the ice sheet's surface.

We can picture the ice bulging out of eastern Canada in all directions, like pancake batter spreading unevenly as it is poured onto a griddle (fig. 6.6). As the ice sheet inched its way into northcentral United States no large obstacle blocked its way, and it sagged into existing lowlands. Where lowlands were large and elongated, as in the case of major river valleys, and particularly where they were aligned in the direction of ice flow, the ice formed into lobes, pressing along the channels of least resistance. In such lowlands the ice's erosive power was especially great. The ice was probably thicker there than over higher land, and moving faster than where it met landform barriers, and generally bedrock there was relatively weak—the reason for the lowlands. As the lobes pushed through those low areas, eroding under the pressure of great weight, they

2. If the ice in Greenland, Alaska, the glacier complex of mountainous western Canada and northwestern contiguous United States, and separate smaller mountain areas is added, the ice-covered area in North America was a third larger than that of Antarctica today.

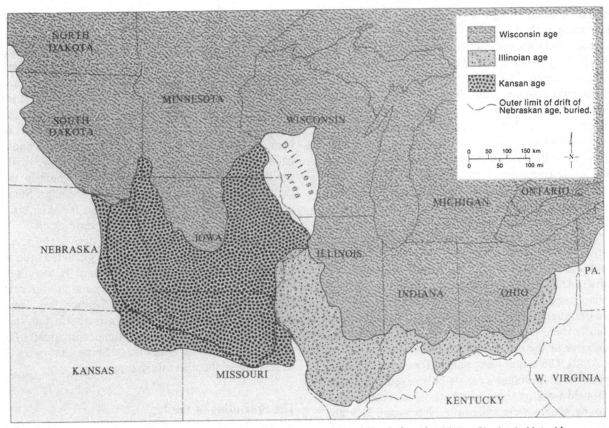

Figure 6.5 The general outlines of the four main drift sheets in central North America. (After Charles L. Matsch)

deepened them. Many lowlands were deepened more with each successive glacial stage: the ice came again and again, reoccupying those lowlands, grinding them ever lower as it slid through them for countless years. Yet, ultimately the lowlands were partly refilled by drift released by the ice.

Some lowlands were deepened so much by repeated glaciation that they became the basins of the Great Lakes. The shapes of lobes that came from the north and northeast can be seen in outlines of the Great Lakes and their larger bays (fig. 6.7).

Several lowlands led the ice into and around Wisconsin. They had been stream-eroded in the weaker of the Paleozoic rocks, and in the sedimentary rocks of the Lake Superior basin, which were weaker than the older Precambrian crystalline rocks flanking the basin on either side. During the last glaciation Wisconsin lay in the path of several powerful lobes. Some halted in northern Wisconsin, slowed by the crystalline-rock highland, and then melted back. Some continued southward—on the east through the weak-rock Lake Michigan basin into Illinois, and on the west into Minnesota. Ice coming across Minnesota into Iowa stopped short of southern Wisconsin, halting west of the Mississippi. Farther north, the tip of one small sublobe from Minnesota did reach a few miles into Wisconsin near Grantsburg, Burnett County (fig. 6.7).

The base of the lobes moving through the Lake Michigan and Lake Superior basins dug to below-sea-

Figure 6.6 Barnes Ice Cap on Baffin Island illustrates how a large spreading glacier develops lobes. Meltwater flows from the ice cap. (Jack D. Ives)

144

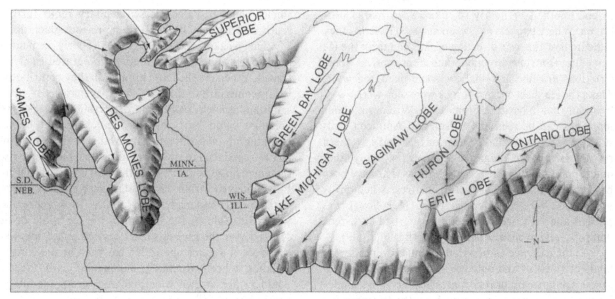

Figure 6.7 The main ice lobes in the Great Lakes region during the Wisconsin Stage of glaciation. Arrows show the general direction of ice movement. Older drift is not shown. (After Charles L. Matsch)

level depths, while the lobes' upper surfaces stood high like plateaus with sloping sides, abutting against neighboring lobes, and deeply burying plains and hills.

Under the ice sheet's tremendous weight the earth's crust was depressed, especially in and around the Hudson Bay area. There the ice lasted longest. All of northeastern North America tilted down in that direction. Since the ice melted off, the lowered crust has risen about two-thirds of the way to its preglacial elevation, and it continues to rebound slowly in the northeast. In the Great Lakes region the rate of rebound now seems to be about 1 foot (0.3 m) every hundred years. As a result, the waters of Lake Superior are gradually, imperceptibly receding from the rising northeastern shores and encroaching on the lake's southern and southwestern shores. There the mouths of rivers flowing into the lake from the south are drowned, and marshes are expanding.

At the time that the Pleistocene ice sheets and other large glaciers existed throughout the world, a significant amount of the planet's water was stored on land, and sea level dropped proportionately. The maximum lowering, according to some estimates, was on the order of 450 feet (140 m), making sea level about 300 feet lower than it is today. Land bridges then emerged connecting Alaska and Siberia, and many other now-separated landmasses and islands. Animals and prehistoric people could migrate and intermingle more freely than they could have today. Rivers eroded their valleys more deeply because their base level, the sea, was lower.

Erosion by Glaciers

A glacier, like other agents of gradation (water and wind), is both eroder and depositor, working to wear down high features and fill in depressions.

As an eroder, a glacier works in various ways. It creeps over the land, pushing and dragging loose material like a mechanical bulldozer and grader. It incorporates soil and rock fragments in its base and carries them along as it moves. It topples weak rock structures like crags, wears down low obstacles, and scours the sides and tops of larger, more durable barriers like bedrock hills.

Glacier ice is plastic, capable of changing shape. The ice and its meltwater penetrate openings in bedrock. As the meltwater refreezes it expands, cracking and prying loose parts of the rock, and holding tightly the loose pieces it has surrounded. When the body of ice slides ahead it plucks up the loose pieces of rock it has gripped and takes them along. Ice's ability to attach itself firmly to objects can be felt as one handles a metal ice-cube tray or piece of ice just taken out of a freezer.

Glacier ice can be exceptionally cold, much colder than 32°F (0°C), which is just the highest temperature ice can have without melting. Very cold ice is harder than ordinary ice, and its grip can be extremely strong. Rocks and grit in its base serve as a giant rasp, which abrades the surface it crosses. While working like coarse-grained sandpaper, the flexible ice molds itself to the land's contours, smoothing them as it keeps passing over.

Ice sheets are the most powerful force ever to cross the land. With great weight from above and pressure from behind, their eroding power is tremendous.

Sharp, resistant rocks held firmly in the base of the ice act as gouging tools, and often cut *striations* (straight scratches and grooves) in the bedrock outcrops over which they are dragged. Those made simultaneously by a certain passage of the ice are parallel to one another, or nearly so, like sweep marks of a broom. Where two sets of "sweep marks" crisscross they indicate how the ice's direction of flow changed; the ice came first from one direction, then from another. Some striations are fine and shallow, some deep and wide. Those the ice sheet cut in soft rock were soon worn away, but many cut in hard rock remain. In Wisconsin, scour markings can be seen at Pattison and Amnicon Falls state parks along the Lake Superior Lowland's southern rim, on parts of the Niagara Cuesta, and on many other outcroppings of durable rock. Striations that have been protected by overlying drift still look fresh when newly uncovered. Where pulverized rock (rock flour) was present under the ice, it sometimes polished hard-rock outcrops, giving them a shiny finish (figs. 6.8, 6.9).

As the ice traveled hundreds of miles over different kinds of surfaces and outcrops, it picked up a heterogeneous assortment of rock materials, including sundry igneous and metamorphic rocks from the Precambrian shield, younger Paleozoic sedimentary rocks farther south, and assorted drift left from previous glaciations.

The weaker rocks like shales, poorly cemented sandstones, and soft limestones disintegrated easily in transit. When crushed and ground up they contributed in large measure to the drift's smaller materials—silts, sands and gravels. But the harder crystalline rocks and dolomite could be transported much farther before breaking up or disintegrating. Many of these eventually were reduced to small fragments too, but many also remained large boulders until the end. Unless they were fully protected within the ice, abrasion rounded off their sharp corners and made them smaller. Those rocks that were held rigid at the bottom of the ice while being scraped over the glacier's bed acquired a flat side, like the facet of a gemstone which has been held against a grinding wheel. Rounded, subangular, and faceted boulders are strewn over glaciated areas, and Wisconsin has its full share of them.

Glacial Erratics

Rocks, of whatever size, that have been glacier-carried from their bedrock source to another location underlain by different bedrock are known as *erratics*. Those with special characteristics can be traced back to their source, and, like striations, are a means of telling in what direction the ice traveled. The most obvious erratics are boulders.

Erratic boulders strewn in the lee of an outcrop constitute a *boulder train*. They are usually distributed over a fan-shaped area, and those dropped close to the outcrop are generally larger and less worn than those carried farther from it. A good example of a boulder train in Wisconsin is that of the quartzite outcrop at Waterloo. There quartzite boulders of the same composition as that outcrop were spread over a fan-shaped area as far as 60 miles (100 km) southwest of the outcrop, showing that the ice had moved over the outcrop from northeast to southwest, varying slightly in direction. Another example of a boulder train is that emanating from Powers Bluff in Wood County. Thousands of boulders of that monadnock's metamorphic rock are strewn for miles to the southeast.

While much glacial material was moved only a short distance, some was carried hundreds of miles. Drift usually consists of a preponderance of local material. For example, in the drift of northern Wisconsin one sees erratic boulders of granite, gabbro, basalt, gneiss, schist, quartzite, iron-formation, and red sandstone, while in southeastern Wisconsin fewer of the boulders are of shield rocks and a greater proportion are of dolomite. The large erratics that were glacier-borne for long distances are the more resistant ones, capable of withstanding abrasion and crushing during transport. Many granitic, gneissic and porphyritic rocks were brought to Wisconsin from Canada (figs. 6.10, 6.11, 6.12).

Figure 6.8 Striations were scratched in this Niagara dolomite when the glacier passed over. They are still fresh, having been covered by drift until now. At the quarry at Valders, Manitowoc County. (Gwen Schultz)

Figure 6.9 Glacial polish on highly resistant granite at Montello. (Lorenz Heim)

Figure 6.10 Impressive erratic boulders often have commemorative plaques mounted on them. This one is Chamberlin Rock on Observatory Hill on the Madison campus of the University of Wisconsin. The book's author, Gwen Schultz, stands before it. Its smooth right side has been ground flat like the facet of a jewel. (Gwen Schultz)

a

b

c

d

e

Figure 6.11 Erratic boulders of Wisconsin. **(a)** Loyal Durand (preparer of Plate 4, Physiographic Diagram of Wisconsin) on extremely bouldery ground moraine, southeastern Marathon County, 1926. (F. T. Thwaites, Wisconsin Geological and Natural History Survey) **(b)** Loyal Durand on a large granite-porphyry boulder. Southeastern Marathon County, 1926. (F. T. Thwaites, Wisconsin Geological and Natural History Survey) **(c)** F. T. Thwaites alongside an erratic boulder, in the late 1920s. (Wisconsin Geological and Natural History Survey) **(d)** Uncleared pasture in eastcentral Portage County in the 1920s. (F. T. Thwaites, Wisconsin Geological and Natural History Survey) **(e)** Boulder fence in eastcentral Portage County. (F. T. Thwaites, Wisconsin Geological and Natural History Survey)

147

Figure 6.12 A glacial erratic being pushed during the construction of Interstate Highway 94 in Waukesha County, 1959. (State Historical Society of Wisconsin)

Among the more distinctive erratics in Wisconsin are fragments and boulders of native copper eroded from the Lake Superior area. They were scattered widely through the Great Lakes region, but were especially numerous near their main sources, two of which were the Keweenaw Peninsula of Upper Michigan and Isle Royale in western Lake Superior. Historians have said that no region received more loose "float" copper than did Wisconsin.

Indians of the Upper Great Lakes region were the first metal fabricators in the Americas. They found pieces of "drift copper" lying on the ground, mixed in the rest of the glacial drift. This copper was in an almost pure state—99 percent pure—and therefore relatively soft, and easy to shape by heating and hammering and by grinding. As the Indians realized copper's usefulness for implements and its beauty for ornaments, and as copper became a trade item, they needed more of it and also dug it from bedrock at the surface (fig. 6.13).

Loose copper erratics worthy of note generally ranged in weight from a few ounces to 50 pounds, but some weighed hundreds of pounds. It is reported that in 1840 a copper boulder weighing 1,700 pounds was found in the channel of the Sioux River on Bayfield Peninsula, 6 miles from Lake Superior. It was sold and removed. An 800-pound mass of copper also was deposited on that peninsula. Copper boulders weighing up to 487 pounds were carried to southeastern Wisconsin. Copper fragments are still found frequently in Wisconsin. Within recent years a 2,200-pound copper erratic shaped like a potato chip was found northeast of Cameron, near the Barron-Rusk county line.

Diamonds are among the erratics deposited in Wisconsin and nearby states, but finding one is a rare event. Several of the diamonds discovered in Wisconsin were outstanding specimens. The first was found accidentally in 1876 while a well was being drilled on a farm near Eagle in Waukesha County. It was known as the Eagle Diamond, and was yellowish and weighed more than 15 carats. In 1880 several small diamond crystals were taken from the banks of Plum Creek, Pierce County, by prospectors panning and sluicing for gold. The largest was three-fourths of a carat. In 1881, a 6.57-carat diamond, white and clear, was discovered north of Saukville, Ozaukee County. A shiny 3.87-carat diamond was picked up by a boy on a farm south of Madison in 1893. A 2.11-carat crystal with a bluish-green tinge was found at Burlington, Racine County, in 1903. Both the Burlington and Saukville stones lacked the frosty surfaces that well-abraded stones have. Other diamonds weighing several carats were found in railroad grade cuttings near Mukwonago in Waukesha County, and in a gravel street in Racine. Wisconsin's largest diamond was picked up near Theresa, eastern Dodge County, in 1888. It weighed 21.5 carats and was almost spherical. One side was colorless, the other almost a cream color. When the Theresa Diamond was sold it was cut into ten individual gem stones.

Whereas some of the diamonds were detected because they were "bright pebbles" or flashed in the sunlight, others looked like ordinary "stones"—dull and irregularly shaped. In recent years small diamonds were reportedly found elsewhere in Wisconsin, but the finds have not been verified. Undoubtedly there are still undiscovered diamonds in the drift.

The much-searched-for source of Wisconsin's diamonds remains a mystery. Since they were eroded and carried by southward-moving glaciers, they came from somewhere to the north—perhaps Canada north of Lake Superior, Upper Michigan, and/or northern Wisconsin. Being the hardest of all natural substances, diamonds are capable of being transported great distances with relatively little reduction in size. Diamonds formed in kimberlite, an intrusive volcanic rock, but not all kimberlites are diamond-bearing. Bodies of it have been located in that northern area, but the source of the glacier-transported diamonds (if it still exists) has not been found.

The Ice Sheet As a Depositor

Eventually all of the ice sheet's massive load had to be dropped, and the way it was laid down created various landform features peculiar to areas of glacial deposition.

Drift is the general, all-inclusive term for anything dropped by a glacier or its meltwater—anything sorted or unsorted, large or small, in any form. Different drift features have special names which will be mentioned as they are discussed. The word "drift" came to be used in Europe before the phenomenon of continental glaciation was even imagined. Then it was believed that this imported mixture of rock debris had been spread over the surface by the flood of Noah's time—that it had "drifted" from the sea onto the land.

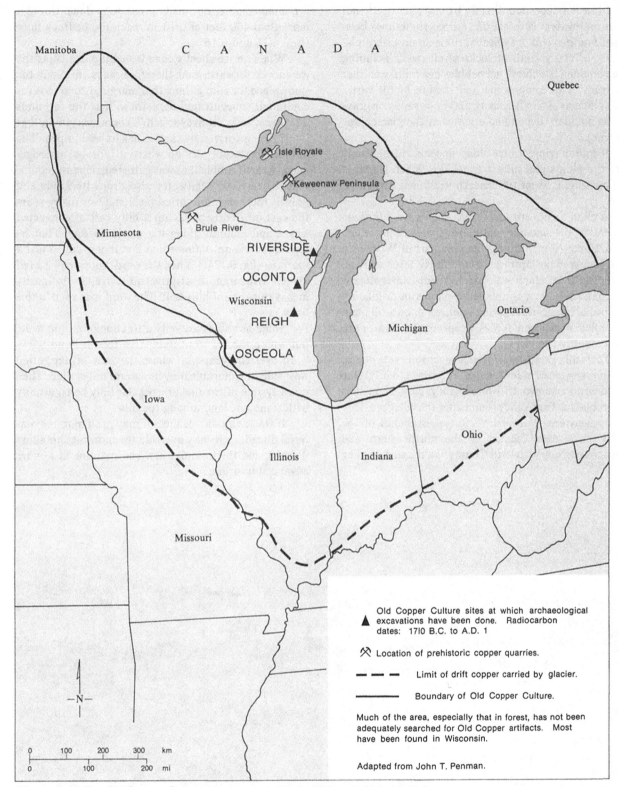

Old Copper Culture sites at which archaeological
▲ excavations have been done. Radiocarbon
dates: 1710 B.C. to A.D. 1

⚒ Location of prehistoric copper quarries.

– – – Limit of drift copper carried by glacier.

——— Boundary of Old Copper Culture.

Much of the area, especially that in forest, has not been
adequately searched for Old Copper artifacts. Most
have been found in Wisconsin.

Adapted from John T. Penman.

Figure 6.13 The Old Copper Culture. (By permission of the *Wisconsin Archaeologist*)

Material deposited directly by a glacier itself (not by its meltwater) is called *till*. (It has sometimes been called *boulder clay*.) Typically, till is an unsorted miscellany of many kinds of rocks of all sizes, including large boulders, cobbles, and pebbles in a matrix of finer materials. The composition and texture of till varies from place to place. It is more clayey or sandy or gravelly or bouldery depending upon what the glacier was carrying.

While moving, an ice sheet plasters till unevenly over the plains and hills it traverses. When stagnant and melting, it drops till beneath itself as it loosens its hold.

A plain with a surface of till is known as a *till plain*. Generally it is smoothly undulating, with relief of less than 20 feet (6 m). In the glaciated part of Wisconsin such plains make up most of the fairly level surface, excluding that which was covered with water-deposited material—for example, outwash plains or glacial-lake beds. The gentle rises and depressions on a till plain are called *swells* and *swales*, respectively. Swales are often marshy (fig. 6.14).

Drift fills old valleys, often to a considerable depth, and in many cases hides a valley's presence totally. One of the largest known drift-filled valleys in Wisconsin is the preglacial Troy Valley (named for the village of Troy in northeastern Walworth County). Branches of the valley have been located in the northeastern and southern parts of Walworth County. On the surface they are undetectable, but drilling has gone down through more than 400 feet of drift to reach the bedrock floor (figs. 6.15 and 6.16).

When an ice sheet's edge is melting as fast as the ice moves forward, and therefore stays in about the same place for a long time, that marginal zone receives continual, concentrated deposition. It is the terminus of a great, wide "conveyor belt," where everything that is still being carried by the ice has to be dumped. The dumped material piles up where it falls—a heterogeneity of earth rubble that was gathered along the course the ice traveled. Meltwater also drops its sands and gravels there in more sorted fashion. Over many years the cast-off debris builds up a hilly belt of connected ridges and mounds along the glacier's edge. That irregular belt is an *end moraine*, commonly called just a *moraine* (fig. 6.17). (When the word "moraine" is used alone, without a descriptive adjective, it ordinarily means this type of hilly belt. The word is so used in this book.)

Some moraines are only a fraction of a mile wide, but where the ice front fluctuated forward and backward over the years, or where the zone of deposition was wide, the moraine may be several miles wide. Then it is a system of parallel or irregular hilly belts, usually with some low land among the hills.

If the ice again advances it may push moraine material ahead; or it may override the moraine, breaking it down, and then build a new end moraine at a more advanced position.

Figure 6.14 An undulating till plain southwest of Verona, Dane County. (B-Wolfgang Hoffmann)

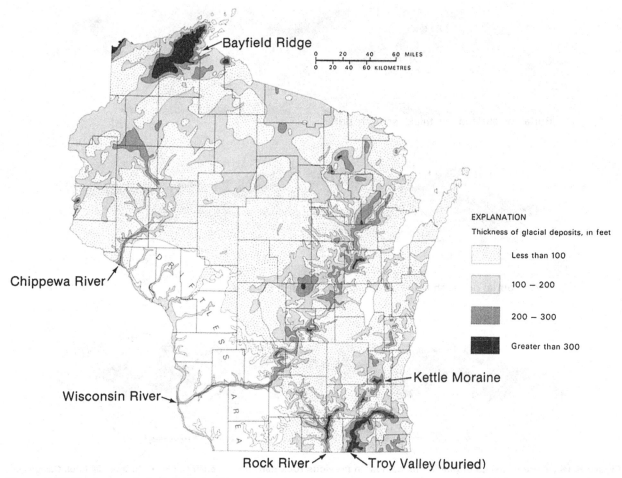

Figure 6.15 Thickness of glacial drift in Wisconsin, generalized. Compare with Plate 2. Notice that some of the places of greatest thickness of drift are river valleys and moraines, especially interlobate moraines. (Wisconsin Geological and Natural History Survey)

The outermost end moraine of a major ice advance, marking the ice's most advanced position is the *terminal moraine*. If the ice pauses while receding, it may build another moraine, called a *recessional moraine,* behind the terminal moraine (fig. 6.18). (Terminal moraines and recessional moraines may also be called end moraines since they form at the outer ridge of an actively flowing glacier.)

Ground moraine is the fairly smooth sheet of debris laid down beneath a glacier like a carpet. Some is deposited while the glacier is moving. If the glacier becomes stagnant—thinning and melting away in place—whatever it carries in its base and within and upon itself is released and added to the previously deposited ground moraine. Till plains consist of ground moraine.

Terminal moraines are more continuous and usually higher than recessional moraines, rising as much as 200 feet (60 m) above the general terrain. However, many are lower than that; and some, or parts of some, are barely identifiable. Moraines made mainly of coarse

materials usually stand higher and firmer than those made of finer materials, which tend to slump and wash away. In the thousands of years since Wisconsin's moraines were built, their contours have been softened by surface weathering, settling, erosion, soil development, soil creep, and vegetation; but those composed mainly of coarse materials retain steep slopes and bumpy surfaces.

By following a moraine, one can trace the outline of the ice front along which it was deposited. Wisconsin's landscape is festooned with moraines (plate 2). Many are striking features stretching across the countryside. They can be recognized as belts of closely set knobby hills and hollows, usually tree-covered and uncultivated because of their roughness and stoniness, However, some are smooth, broad-topped ridges. If the moraines are in their natural, undisturbed state, erratic boulders can usually be seen on and near them.

Some of the individual hills within a moraine are *kames*—steep-sided mounds of water-sorted sands and

151

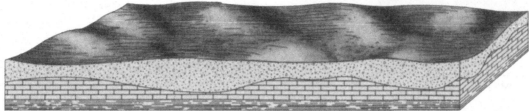

A. Burial of surface by thick, smooth drift.

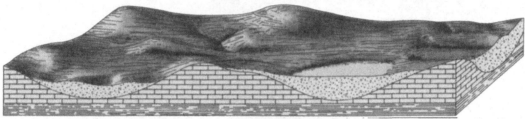

B. Partial burial of rough land—a rock-controlled drift surface.

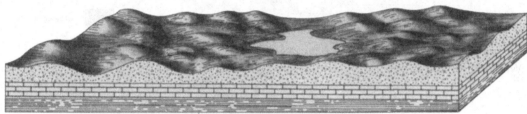

C. Burial of smooth surface by rough moraine.

Figure 6.16 Different results of glacial deposition on previous rock surfaces. (Vernor C. Finch. McGraw-Hill Book Company. *Elements of Geography*, 5th ed., 1967, by Trewartha, et al.)

gravels that were built where streams of meltwater draining off stagnant ice, or draining down holes through it, dropped their loads as their velocity suddenly decreased. Kames were built back from moraines also, within or alongside a wasting glacier. Kames that formed at the bottom of round holes in the ice may be symmetrically conical in shape. Others that accumulated like steep deltas or fans against the ice's edge or in crevasses, and that slumped when the ice was gone, are irregular in shape (figs. 6.19, 6.20).

Hollows on the surface of moraines are called *kettles* (or *kettle holes*). Kettles are not confined to moraines, but are found on other surfaces of glacial deposition as well. Some are the result merely of uneven deposition. Others were caused by large blocks of ice left detached from the main body of the glacier as it disintegrated. Where those ice remnants stood, material could not be deposited. Some ice remnants were deeply buried under drift, well insulated from summer warmth. They melted slowly, large ones lasting perhaps for hundreds of years. As they gradually disappeared, the drift around and above them sank or washed into the vacated space, partly filling it, but at the surface a distinct depression remained where an ice rem-

nant had been. Such a feature is also called an *ice-block depression*. Many such depressions are round, but most are irregular. Some hold a pond or marsh because they have no stream outlet; water seeps away through the porous drift. Some large ones contain lakes. (Figs. 6.21, 6.22.)

Because moraines are composed of small hills and hollows bunched together, they have no organized drainage pattern. A large moraine is usually a drainage divide.

Moraines, with their kame-and-kettle topography, are often described as a jumble of hills, a hummocky hodgepodge of glacial rubble, and in a sense they are. But people who carefully examine their structure can learn the sequence in which a moraine's individual features were constructed, and analyze what the ice was doing in that locality. Wisconsin geographer Adam Cahow, a specialist in glacial landforms, had that perception and often expressed it to participants on field trips, pointing out, "A moraine is not a chaos of hills, but an orderly assemblage of specific landforms." Glacial geologists today are reading moraines in that light, piecing together the details and chronology of glacial happenings.

Figure 6.17 A moraine being built in the foreground by a valley glacier in Mount Cook National Park, New Zealand. Here material of mixed size and composition is being dropped haphazardly, though some sorting is being done by meltwater draining away. Large blocks of ice remain partly covered by rock debris that the glacier carried. (New Zealand National Publicity)

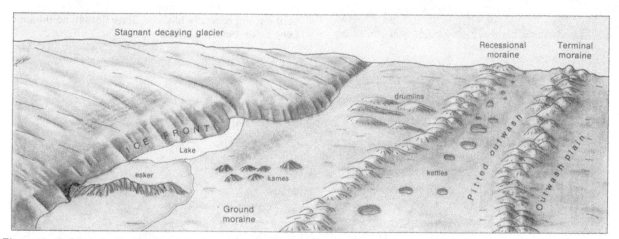

Figure 6.18 Some features left by a retreating, withering ice sheet. The ice front is melting backward to the left. The terminal moraine is along the line of the ice front's farthest advance. The recessional moraine is where it paused during retreat. (Gwen Schultz)

153

Figure 6.19 Mounds like these accumulating against the edge of a stagnant, melting Canadian glacier, built by streams draining off the glacier's surface, illustrate how some kames form. (Gwen Schultz)

a

Figure 6.20 (a) The largest kame in the northern unit of Kettle Moraine State Forest, near Dundee—250 feet high. (George Knudsen, Wisconsin Department of Natural Resources) **(b)** Looking north at McMullen Hill at left and Connor Hill at right, examples of conical kames, which rise more than 100 feet above the flat plain. In the northern unit of Kettle Moraine State Forest, northeast of Long Lake. (Robert F. Black)

b

Figure 6.21 A kettle hole alongside a moraine near Prairie du Sac holds a small pond. (Gwen Schultz)

Figure 6.22 Butler Lake in the northern unit of the Kettle Moraine State Forest occupies a kettle. (George Knudsen, Wisconsin Department of Natural Resources)

The most prominent end moraines in Wisconsin, and across the country, are those left by the last major ice invasion, which occurred late in the Wisconsin Stage. (Locally this Late Wisconsin advance of the ice sheet was formerly called "Woodfordian," a term derived from Woodford County in central Illinois.) (Plates 2 and 4, and fig. 6.2.)

The Late Wisconsin terminal moraine arcing across Wisconsin did not form simultaneously along its entire length, for the ice did not reach or withdraw from its outer limit in all places at the same time. Where this moraine enters St. Croix County from Minnesota, it formed an estimated 15,000 years ago. Sections of the moraine farther to the east and south in Wisconsin seem to have formed later, about 14,000 years ago.

Several ice lobes built this Late Wisconsin system of terminal moraines in Wisconsin. In the northwest two large lobes covered northern Wisconsin in Late Wisconsin time—the Superior Lobe and the Chippewa Lobe. The next large lobe to the east was the Green Bay Lobe, which bulged into southern Wisconsin. East of it was the Lake Michigan Lobe, which occupied the Lake Michigan basin. As the Lake Michigan Lobe grew, its outline was similar to that of the lake basin it scooped out, though the lobe spilled over into eastern Wisconsin and western Michigan. At its fullest extent it spread south into Illinois and Indiana (fig. 6.5, 6.7).

The Lobes of Northern Wisconsin

The Superior and Chippewa lobes took shape when ice coming from the north was divided by Bayfield Peninsula. The ice passing and overriding the peninsula scoured it and sculptured the group of hills that, when partly submerged, would become the Apostle Islands. The moraines built by the Superior and Chippewa lobes are some of the highest and roughest in Wisconsin (fig. 6.23 and plates 2, 3, and 4).

The Superior Lobe, the ice west of Bayfield Peninsula, slid southwest through what is now the Lake Superior basin. Directed by topography, it followed an old river-worn lowland of relatively weak sedimentary rocks, eroding deeper than any river could. Lake Superior's water surface now is about 600 feet (183 m) above sea level, but its maximum depth is 1,333 feet, of which 727 feet (222 m) are below sea level. Because of the lake's great depth its water stays cold. In late summer the surface water has warmed to about 70° Fahrenheit (21°C) but water at a depth of 300 feet is only about 40°F (4°C). The Superior Lobe's terminal moraine, called the St. Croix Moraine, extends through St. Croix County into Minnesota.

Along the east side of Bayfield Peninsula moved the Chippewa Lobe, shaping Chequamegon Bay, and advancing over Wisconsin's northern dome and part way down the Chippewa River valley into what is now Chippewa County. To the southwest this lobe was slowed where it met the quartzitic Blue Hills, and only partly overrode them.

The Chippewa Moraine, built by the Chippewa Lobe, contains a great variety of well-preserved glacial features. Most of it is hummocky with steep-sided hills separating hollows. The hills and hollows generally have a relief of 20 to 100 feet. Many hollows contain small lakes or marshes. A section of the Chippewa Moraine near Bloomer in Chippewa County is included in the Ice Age National Scientific Reserve.

The Chippewa Moraine is one of Wisconsin's best examples of a "dead-ice" moraine. As the Chippewa Lobe withdrew it left at its outer margin a morainal belt several miles wide containing stagnant, "dead," drift-covered ice. Some masses of ice buried under the insulating blanket were more than a hundred—perhaps several hundred—feet thick, and probably did not completely melt for centuries.

Large dead-ice masses melted unevenly, and within some there came to be low ice-free areas that were still surrounded by higher ice. In these ice-enclosed depressions small lakes collected, and over the years the lakes

Figure 6.23 The last major ice advance of the Wisconsin Stage—the glacial lobes and drainage. (Lawrence Martin, *The Physical Geography of Wisconsin,* 1965, with permission of The University of Wisconsin Press)

deposited fine sediments that formed a relatively flat floor. When the encircling ice broke up, the lakes drained away leaving their old floors, which were then usually higher than their surroundings. These former lake floors are termed *ice-walled-lake plains.* There are many of them in the Chippewa Moraine. Characteristically they are flat-topped hills whose surfaces are smooth areas of cultivated land, up to a mile across, within the otherwise rough morainal belt of less usable, stony terrain (fig. 6.24).

The highest point in Wisconsin—Timm's Hill, 1,951.8 feet (595 m) above sea level—is part of the Chippewa Moraine where it curves northward, several miles east of Ogema in southeastern Price County (fig. 6.25). Nearby is the state's second-highest point, Pearson Hill (1,950.8 feet), also part of the moraine. One reason for the superior altitude of these hills is the fact that the moraine in that area lies on the already-high northern dome.

U.S. Highway 53 runs alongside the Chippewa Moraine, a few miles west of it, from north of Chippewa Falls northwest to Bloomer and Rice Lake. At the border of Barron and Washburn counties, the highway is at the juncture of the moraines left by the Chippewa and Superior lobes (plate 2).

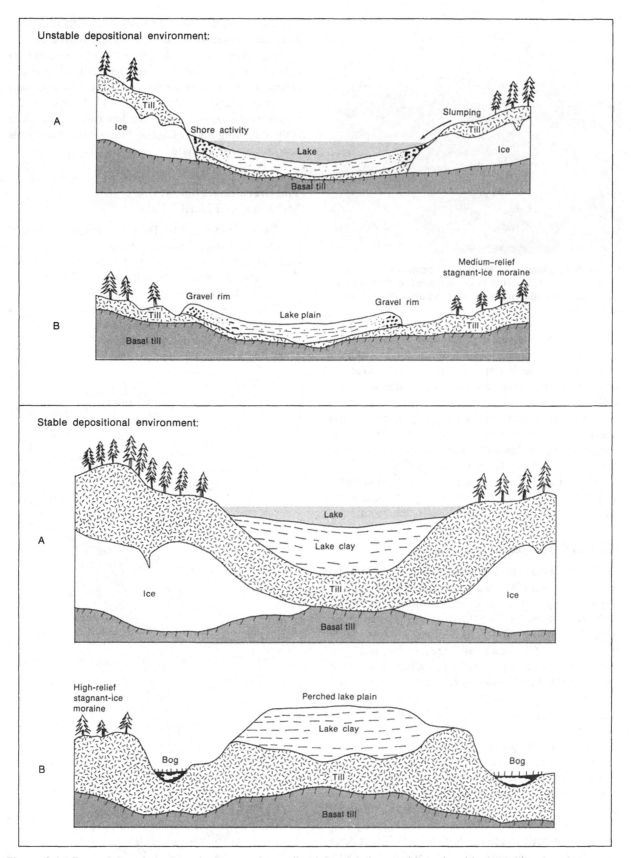

Figure 6.24 Formation and characteristics of an ice-walled-lake plain in unstable and stable depositional environments: A—with ice present; B—after ice melted away. (Lee Clayton)

157

Figure 6.25 Timm's Hill, the highest point in Wisconsin—1,952 feet (595 m), a mound of glacial drift upon the Precambrian shield, southeastern Price County. (Dan Hein)

Along the line where these two lobes met and flowed next to each other, a composite moraine was built between them. Such a moraine, between two lobes, is known as an *interlobate moraine*. This is one of several in Wisconsin.

To the east, between the Chippewa Lobe and the Green Bay Lobe, were two minor lobes. One was the *Wisconsin Valley Lobe* which moved into the valley of the upper Wisconsin River in Lincoln County. Immediately to the east of that lobe the *Langlade Lobe* came from the northeast, reaching just into Langlade County. Both lobes left moraines that are short but prominent.

The Northern Zone of Old Drift

South of the curving moraines of northern Wisconsin described above is a wide zone of older drift. It was deposited before the Late Wisconsin ice advance built those moraines. One traverses this old-drift zone when driving west from the Wausau area on State Highway 29 to Chippewa Falls, Menomonie, and Prescott on the Mississippi River. The indistinct southern limit of this old drift runs approximately (east to west) from south of Stevens Point to north of Wisconsin Rapids, west across southern Clark and Eau Claire counties, and then south through Pepin County to the Mississippi River (plates 2 and 4).

The old drift is thinner, less continuous, and more weathered and eroded than the newer Late Wisconsin drift to the north. In that its glacial characteristics have become less perceptible with the passage of time, the old drift is transitional between the fresh, easily recognized new drift to the north and the Driftless Area to the south. Consequently its border with the Driftless Area through this region is not well defined.

A corridor along the Wisconsin River, about 10 miles (16 km) on both sides of the river, extending to

north of Wausau, used to be considered driftless and was mapped as a northward-extending panhandle of the Driftless Area. But in Marathon county there are erratics from far to the north and remnants of deep till, traces of an old drift cover that has been largely removed by erosion. It does seem unlikely that the ice would have always avoided the low river valley while fully occupying higher land on both sides. This area is now considered glaciated.

The Green Bay Lobe, Its Moraines, and Devil's Lake

Door Peninsula, like Bayfield Peninsula, divided the south-flowing ice into two lobes—the Lake Michigan Lobe and its western offshoot, the Green Bay Lobe. That wedge of resistant dolomite was severely abraded, but endured repeated glaciations.

The *Green Bay Lobe* developed asymmetrically as it lengthened southward, spreading more to the west of its axial growth line than to the east (plate 2 and fig. 6.26). On its west side, where free to expand, it wid-

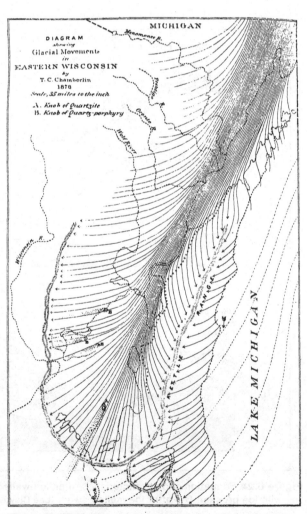

Figure 6.26 Diagram showing glacial movements in the Green Bay and Lake Michigan lobes. (T. C. Chamberlin)

158

ened to the center of the state, but on the east it abutted against the Niagara Cuesta, which it overrode, and against the larger Lake Michigan Lobe. The Green Bay and Lake Michigan lobes advanced along open north-south lowlands, growing along parallel paths beside each other. They were not retarded as other lobes to the northwest were by Wisconsin's northern highland.

The Green Bay Lobe followed the belt of soft Paleozoic rocks west of the Niagara Cuesta, scraping them out and creating depressions later occupied by the waters of Green Bay, Lake Winnebago and Horicon Marsh. The scouring-out of soft sandstone excavated the basin of Green Lake, Wisconsin's deepest inland lake, whose depth is about 230 feet (70 m). The moving ice cleaned away spurs and consumed outliers of the Niagara Cuesta, smoothing its outline and planing back parts of its escarpment perhaps as much as five or ten miles; and it advanced into the upper valley of the Rock River (plates 3 and 4).

The terminal moraine of the Green Bay Lobe is nearly continuous, outlining the lobe's maximum proportions. On the south and southwest sides of the lobe it is known as the *Johnstown Moraine*. Its southernmost part in Rock County is near Johnstown. Westward from Johnstown the moraine crosses the Rock River north of Janesville, and curves past Verona and Cross Plains toward Prairie du Sac (fig. 6.27).

A section of this moraine at Cross Plains is a unit of the Ice Age National Scientific Reserve (fig. 6.1). At that site the moraine is relatively small, but there are other interesting features, including the remains of small lakes and drainageways which were at the glacier's edge, and just a few yards beyond the moraine are sculptured dolomite outcrops that would have been destroyed if the ice had moved that much farther.

This terminal moraine of the Green Bay Lobe is partly obscured in the Wisconsin River Valley, but northward it becomes evident again. It crosses the Baraboo Hills at Devil's Lake (also a unit of the Ice Age National Scientific Reserve); continues past Baraboo, going south and east of the Wisconsin Dells; delineates the eastern margin of the Central Sand Plain; and extends north to meet the Langlade Lobe northeast of Antigo (plate 4).

The Green Bay Lobe left numerous recessional moraines. In the southwest, a few miles behind the Johnstown Moraine and concentric with it, is the well-developed *Milton Moraine* (named for the village of Milton, northeastern Rock County), representing one position where the ice paused during its withdrawal (fig. 6.28). It crosses Madison's west side.

Farther back at a position of further recession is, or was, the smaller *Wingra Moraine*. At least, it was recognized in early maps and literature. It is described as extending through central Madison from the old University of Wisconsin nucleus on Lake Mendota, south-southeast between lakes Wingra and Monona, along the area of Mills and Park streets. Perhaps travelers originally took advantage of slightly higher ground there. Now, however, after much grading, a moraine is hardly recognizable. Modern mapping shows that area as part moraine and part lake-bed sediments.

In a position of still farther retreat is the *Lake Mills Moraine,* a patchy moraine belt. Interstate Highway 94 crosses it west of Lake Mills. And there are other recessional moraines. Farther north many parallel north-south moraine belts lie behind the Green Bay Lobe's terminal moraine; they were built as the ice shrank, sometimes readvanced a little, and then shrank still farther.

Figure 6.27 Part of the Johnstown Moraine near Cross Plains, western Dane County. Notice the mixed composition of the drift exposed in a road cut. (Robert F. Black)

Figure 6.28 North face of the Milton Moraine, a mile south of Koshkonong, southwestern Jefferson County, in 1938. The stream is Otter Creek. (Milwaukee Public Museum. Photographer, Kenneth Vaillancourt)

Devil's Lake owes its existence to the Green Bay Lobe's terminal moraine (figs. 6.29, 6.30). Apparently the preglacial Wisconsin River used to flow south through the Lower Narrows in the North Range of the Baraboo Hills, and on through the Devil's Lake gorge in the South Range, which was deeper than it is now. The Green Bay Lobe pushed into the Wisconsin River Valley, acting as a dam and forcing the river to abandon its old course through the Baraboo Hills. The ice engulfed the east end of the Baraboo Hills but was slowed by the high South Range, stopping on its top before reaching the gorge. It moved more easily over lower ground, however, and crept along the base of the range on both sides until it came to the gorge. For many years ice fronts loomed over both ends of the gorge, forming terminal-moraine dams there and spilling forth outwash. *Outwash* is drift (mainly the smaller materials) that is washed out from a glacier by its meltwater. It is usually sorted and stratified by the flowing water (fig. 6.31). The gorge, which is thought to have been all of 900 feet (275 m) deep when the preglacial river flowed through it, became almost half filled with outwash. Quartzite bluffs now rise as much as 500 feet above the gorge floor and Devil's Lake. Within the gorge, where the eroding ice did not reach, talus drapes the bluffs. It is not known over what length of time the blocks of talus accumulated.

When the ice sheet receded the river could not return to its gorge, for the old course was blocked by moraines and other drift. So it took the lowest route, which it still uses—the big bend around the east end of the Baraboo Hills, by the site of Portage. Near Prairie du Sac the river reentered its old valley, going west through the Driftless Area. The tributary Baraboo River had a new course too; it flowed north through the Lower Narrows to meet the Wisconsin River near Portage (fig. 6.29).

Because moraines dammed the Devil's Lake gorge at both ends, a lake formed between them. Devil's Lake now is about 1.3 miles (2 km) long, and 0.4 to 0.6 miles wide; and normally it is 30 to 40 feet (12 m) deep. Two small streams flow into the lake, but it has no surface outlet. Still its water is fresh (not salty) because it seeps out through the porous, gravelly outwash beneath it. The lake floor is at a higher elevation than the till plain beyond the moraine.

The Devil's Lake area is one of the most spectacular geologic sites in the state, and the most visited unit of the Ice Age National Scientific Reserve. It is also a popular camping and recreation spot. Here one sees a condensed record of geologic history from the ancient Precambrian to the Present. In the hills themselves are revealed evidences of Precambrian deposition, metamorphism, rock-folding, and erosion. Paleozoic sedimentary rocks that once completely covered the hills have been partly eroded, letting the hills emerge, and

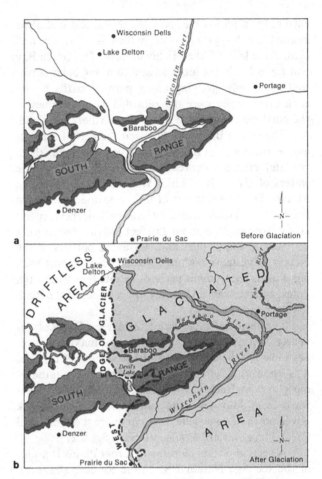

Figure 6.29 (a) Presumed preglacial drainage lines across the Baraboo Hills area. **(b)** Present courses of Wisconsin, Baraboo and Fox rivers. (After Hotchkiss and Bean, 1925)

are slowly being worn off even more. From atop the craggy, purplish quartzite bluffs at Devil's Lake one looks down to where in preglacial times a river used to flow majestically through the gorge, which then was about twice as deep as it is today. From those heights one views two segments of moraine blocking both ends of the gorge, and the lake held between, and envisions the time when ice walls stood where the moraines are now, and an iceberg-dotted lake, which rose higher than present Devil's Lake, occupied the gorge. To the southeast beyond the moraine, one sees the till plain where the ice sheet used to lie, stretching over the horizon; and to the west one sees the Driftless Area without the veneer of drift. A dramatic panorama, with much geology in view.

The Lake Michigan Lobe

In its last major advance the *Lake Michigan Lobe* moved south along the lake-basin zone of weak sedimentary rocks east of the Niagara Cuesta. It and earlier lobes that had passed that way eroded Lake Michigan's bed more than 300 feet (90 m) below sea

Figure 6.30 View from atop the South Range of the Baraboo Hills, looking north across the north end of Devil's Lake and the wooded terminal moraine that blocks that end of the gorge. (Roger Peters)

Figure 6.31 Stratified fine material laid down by glacial meltwater, layer upon layer. South of Newburg, Washington County. (Milwaukee Public Museum. Photographer, George Gaenslen)

level. In its deepest place, southeast of Sturgeon Bay, the lake bed is about 923 feet below the present lake level, which is about 584 feet (178 m) above sea level. Of course, drift was deposited on the bedrock floor of the basin.

This Lake Michigan Lobe extended farther than the present lake, and was about 20 to 40 miles wider than the lake on each side. Its terminal moraine (with parallel recessional moraines) arcs from Michigan through northern Indiana and Illinois into Wisconsin, where it meets the moraine of a minor lobe, the *Delavan Lobe*—in Walworth County. North of that moraine the Lake Michigan and Green Bay lobes together formed the Kettle Moraine, described below.

The Delavan Lobe was a little bulge off the Lake Michigan Lobe, right next to the end of the Green Bay Lobe. It may have been diverted into the preglacial, now-buried Troy Valley (fig. 6.15). It extended far enough to the southwest to cover the Delavan and Lake Geneva area, reaching Darien. Its short, crescent terminal moraine is called the *Darien Moraine*. About five miles behind that is its recessional moraine, the *Elkhorn Moraine,* named after Elkhorn in central Walworth County (plates 2 and 4). Lake Geneva and other nearby lakes, lying among glacial hills, are known for the beauty of their surroundings. The resemblance between Switzerland's Lake Geneva and its namesake in Wisconsin—both long, narrow, and in a glaciated setting—is apparent.

Recessional moraines left by the Lake Michigan Lobe in eastern Wisconsin lie generally parallel to the lakeshore. Trending north-south, they direct the drainage in southeastern Wisconsin. The courses of many streams were set even while the ice was present.

Then water flowed along the ice's margin, which, like the moraines it built, had a general north-south alignment. Some rivers, such as the Milwaukee and the Root, flow behind lakeshore moraines for a long way before being able to break through to the lake only a short distance away.

Old Drift of Southcentral Wisconsin

Old drift covers southcentral Wisconsin south of the Green Bay Lobe and west of the Delavan and Lake Michigan lobes—much of Rock and Green counties and small parts of adjacent Dane, Walworth and Lafayette counties—and extends south into Illinois (plate 4). The older drift predates the lobes mentioned and was left by more than one glaciation. The glacial aspect is missing over much of this old-drift area because of dissection by the Sugar and Rock rivers, the thinness of drift on bedrock topography, and the scarcity of readily visible glacial deposits and features.

The Kettle Moraine

Between the Green Bay and Lake Michigan lobes was built the *Kettle Moraine,* sometimes called the "Kettle Interlobate Moraine" or "Kettle Range" in older literature (plate 4). Actually there are many moraines with kettles in Wisconsin and the world, but this is a classic one. It is of grand proportions: length well over 100 miles (160 km), width from 1 to 10 miles, and relief from 100 to more than 300 feet (90 m). It extends from Kewaunee County south to beyond Whitewater Lake in northwestern Walworth County, almost paralleling the Lake Michigan shore. For countless years along that zone thick ice masses spreading sideways from the two large lobes discarded their double loads of drift and spilled their intermingling torrents of meltwater (fig. 6.32).

Today hiking trails and Kettle Moraine Drive wend their way along that belt of hills, among its woods and kettle lakes. Two sections of Kettle Moraine have been designated state forest units to help protect their natural features. The southern unit extends from Whitewater Lake north into southwestern Waukesha County. The northern unit, in the moraine's midsection, extends from Kewaskum in northern Washington County north through southeastern Fond du Lac County to Glenbeulah in western Sheboygan County; it is included in the Ice Age National Scientific Reserve.

Kettle Moraine contains much locally derived rock material, especially dolomite, as well as erratics from northern Wisconsin, Upper Michigan and Canada. It

Figure 6.32 A section of the Kettle Moraine west of Eagle, Waukesha County. Hummocky surface with kettle holes; the one in the foreground holds a pond. (Milwaukee Public Museum)

is replete with kettles and kames and other glacial features. Here many types of glacial formations were first studied and interpreted in North America.

The highest point in southeastern Wisconsin is a hill on the crest of Kettle Moraine in southwestern Washington County, just south of State Highway 167. There the moraine lies atop the Niagara Cuesta. This high point—1,335 feet (407 m)—is known as Holy Hill because of the monastery on its summit. It was formerly called Lapham's Peak in honor of Increase Lapham, who in 1851 was the first to measure its altitude by instrument, and who reported finding three Indian mounds at its top. More recently another Kettle Moraine hill to the south was named for him: *Lapham Peak,* south of Delafield and Interstate 94, is Waukesha County's highest point—1,233 feet. Formerly it was called Government Hill. It is recognizable by its tall signal tower and shorter observation tower. From the observation tower there is an excellent view of the rambling moraine and some of the lakes it holds among its kames and ridges. Understandably this moraine region is a popular recreation area.

Retreat of the Ice Sheet

The many-lobed Late Wisconsin ice sheet drew back from its moraine-scalloped margin and gradually receded north out of Wisconsin (fig. 6.2). As it vacated the basins that would become the Great Lakes, bodies of water collected in them against the shrinking ice.

These water bodies (composed then of glacial meltwater and runoff from the land) grew progressively larger as the ice relinquished more and more of the basins' area. This was the beginning of the Great Lakes, but they would go through many stages before reaching their present outlines.

The ice sheet did not make a uniform, steady retreat. Its lobes oscillated during their final withdrawal, melting back and readvancing a number of times, so the impounded water bodies that were the predecessors of the Great Lakes were continually changing in size and shape in response to the ice's position (fig. 6.33). The waters of an ice-dammed lake drained through the lowest available outlet, but that was subject to change. If that outlet became blocked, the lake rose until it could overflow elsewhere. As outlets at different elevations were alternately dammed by the ice and reopened as it melted back, the early lakes' levels rose and fell and their outflow spilled first in one direction, then in another. When lake levels were higher than they are now, adjacent lowlands were flooded and some of the Great Lakes merged. While the ice sheet lay across the St. Lawrence River outlet to the Atlantic Ocean or across waterways that now connect the lakes, lake water drained southwestward to the Mississippi River. Even after the ice had left the area, the slow uplifting of the continent in the northeast Hudson Bay region caused drainage to be redirected. Only a few thousand years ago did the Great Lakes come to have their present levels and shapes. Modern fluctuations in lake levels,

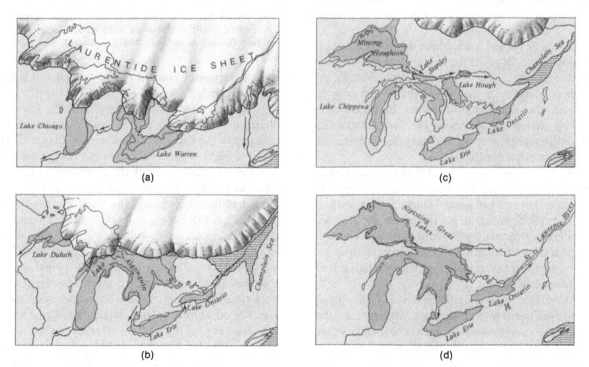

Figure 6.33 Stages in the development of the Great Lakes: **(a)** 13,000 years ago; **(b)** 11,500 years ago; **(c)** 9,500 years ago; **(d)** 6,000 years ago. (After Charles L. Matsch)

though of concern to lakeshore property owners, are quite minor compared to the large fluctuations of the past.

Low, treeless tundra was probably the first vegetation to cover Wisconsin's ice-free land, and as the climate moderated higher types of plant growth followed, including boreal evergreen forest. This hardy forest could grow close to the ice where the climate was not too severe, just as today glaciers and forests exist side by side in some temperate regions. Buried remains of one coniferous forest in eastcentral Wisconsin tell of an interesting episode in the closing phase of the Ice Age in Wisconsin. Their best-known exposure is the Two Creeks Forest Bed.

Events Recorded in the Two Creeks Forest Bed

The *Two Creeks Forest Bed* is exposed in an eroding bluff overlooking Lake Michigan northeast of the village of Two Creeks at the boundary of Manitowoc and Kewaunee counties. This site is a unit of the Ice Age National Scientific Reserve (fig. 6.1).

The Two Creeks buried forest bed is an internationally recognized stratigraphic horizon, or level, because it lies between two different layers of drift—one that the forest grew upon, and one that buried it—and it has supplied dates that are widely referred to as a time gauge. According to radiocarbon dates of forest remains, the forest existed there about 12,200 to 11,400 years ago. It covered inland areas as well. Its remains are seen also in the Green Bay-Fox River lowland. (Similarly preserved interglacial forest beds of younger age have been found too.)

Layers of till and glacial-lake deposits exposed in the lakeshore cliffs relate the sequence of events in this period near the close of the Ice Age. Not all researchers interpret the deposits in the same way or reconstruct the same story of what happened. The following account of events at the Two Creeks site, therefore, is quite general, and may need revision as more becomes known (fig. 6.34).

The till at the base of the cliff at the Two Creeks site—that beneath the forest bed—was laid down by an ice lobe that reached some distance farther south in the Lake Michigan basin. At that time, the lobe was advancing in a temporary forward pulsation which interrupted its gradual retreat. The underlying till it laid down is compact and clayey, with colors ranging from gray to reddish.

Above that underlying till is a layer of laminated lake-bed sediments several feet thick. They indicate that the ice had melted back again, and that high lake waters had covered this site. The ice still blocked the outlet at the lake's north end. The warming trend continued and the ice retreated even more. The ice blockage in the north melted away, the lake level fell, and trees grew

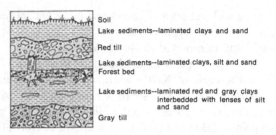

Soil
Lake sediments—laminated clays and sand

Red till

Lake sediments—laminated clays, silt and sand
Forest bed

Lake sediments—laminated red and gray clays
  interbedded with lenses of silt
  and sand

Gray till

Figure 6.34 Generalized cross-section showing the stratigraphy at the Two Creeks site. (After Charles L. Matsch)

Figure 6.35 The dark layer is the interglacial forest-bed horizon of organic material as seen at one location at the Two Creeks exposure. (Robert F. Black)

along the shore of Lake Michigan (which was smaller than it is now).

This warm episode of forest growth, known in Pleistocene chronology as the *Two Creeks Interval,* is recorded in the cliff-face by a buried soil layer of dark brown to black organic matter up to a few inches thick. At different locations along the shore, and at different times as the bluffs erode, a changing assortment of forest-bed materials comes into view at that horizon (figs. 6.35, 6.36). Sooner or later they tumble down the cliff and fresh materials behind become exposed. One sees the continuous disinterment of the immature forest soil and the litter of the forest floor—twigs, remarkably well-preserved branches, small logs from fallen trees (mostly spruce), needles and cones, and the remains of other forest plants. Some rooted tree stumps still stand in growing position. Tree rings show that at least one tree lived as long as 142 years. The forest's life was surely longer than that. While the forest existed the climate there was much like that north of Lake Superior today, where spruce forests are at home.

Figure 6.36 A log protruding from the interglacial forest bed near Two Creeks. As the bluff erodes back from the Lake Michigan shore, remains of the forest bed are exposed. (Charles L. Matsch)

Figure 6.37 The advancing Taku Glacier in the Coast Mountains of southeastern Alaska overruns a forest, illustrating how trees were toppled at Two Creeks and other parts of eastern Wisconsin late in the Ice Age. (William O. Field, American Geographical Society)

But warmth was not to last. The ice began moving south again and reblocked the northern outlet of early Lake Michigan. When the lake rose again the water drained out at a break in the morainal belt near Chicago, through the Des Plaines River to the Mississippi. Flooding of the lakeshore forest and cooling of the climate caused the trees to die. Any trees still standing when the glacier returned were pushed over by it. At the exposed site many broken-off trunks can be seen lying to the lee side of their stumps, where they fell as the glacier overran and flattened the forest (fig. 6.37).

The forest litter was buried under a new layer of lake-bed sediments. And over that the southward-moving ice deposited a layer of red drift several feet thick. Its redness comes from the iron-oxide-rich material picked up in the Lake Superior region. (At other locations along the shore, the layers of drift below and above the forest bed may be different.) It used to be generally agreed that the ice that overran the Two Creeks Forest Bed advanced as far south as Milwaukee, because distinctive red drift like that overlying the forest bed is spread that far. But some geologists now believe that the ice that buried the Two Creeks Forest Bed reached only as far south as Two Rivers, Wisconsin (and Manistee, Michigan, across the lake), and that the red drift farther south was deposited by earlier ice (predating the Two Creeks forest).

Until recently, all red drift of the Late Wisconsin glaciation in eastern Wisconsin was called *Valders till,* named for its type location at Valders west of Manitowoc, about 20 miles southwest of Two Creeks. It is now known that not all of eastern Wisconsin's red drift is of the same age, that not all of it was deposited by the same spreading of ice that deposited red drift at Valders. Some red drift came from the Green Bay Lobe, some from the Lake Michigan Lobe. At this writing, the questions of how widespread or limited the "Valders" till is, and what the relative ages of red drift deposits in various locations are, and what they should be named, are unsettled.

Late Events in the Lake Superior Area

A last readvance of the Superior and Chippewa lobes in northern Wisconsin sent ice southward again into the region of Douglas, Bayfield, Ashland, Iron, and northern Vilas counties, across which they left a massive east-west moraine (plate 2).

Morainal material was left also on Bayfield Peninsula, which divided the lobes. This material forms part of Bayfield Ridge, which consists partly of till but mainly of sandy outwash. This is one of the thickest local amassments of glacial deposits in Wisconsin, in places up to 500 feet thick. The sandy outwash of Bayfield Ridge contained leftover blocks of buried ice; when they melted, empty hollows remained. Bayfield Ridge is highest along its southeastern, Chequamegon Bay side, where the edge of the outwash resembles an escarpment. Mount Ashwabay (a coined word combining Ashland, Washburn and Bayfield) on that side of the peninsula is built of outwash, and has an elevation of 1,320 feet (402 m). Several nearby hills exceed that height. One just west of it reaches 1,435 feet.

As the ice began its final withdrawal from the Lake Superior basin, water was ponded between the glacier and the divide to the south, forming small lakes on

Figure 6.38 Looking over the Duluth-Superior harbor toward Superior, Wisconsin, from a TV tower in Duluth. The harbor is formed by the drowning of the mouth of the St. Louis River which enters from the right, and by the protection of Minnesota Point and Wisconsin Point, the bars at the left. On the horizon is the level surface of the Precambrian shield in Wisconsin. (Donald H. Jackson)

either side of Bayfield Peninsula. As the ice retreated north beyond the tip of the peninsula, those lakes joined to form one larger lake at the western end of the Lake Superior basin. This ancestor of Lake Superior is known as *Glacial Lake Duluth* (fig. 6.33). It grew larger as the ice withdrew and vacated more of the basin. The lake's level rose and fell, and at one time was about 500 feet above the present level in the Superior-Duluth area. Glacial Lake Duluth drained south through the valley of the Bois Brule River, over the present divide and into the St. Croix River, and then on into the Mississippi. The St. Croix Dalles, cut by the water pouring out of this lake, is now part of the Ice Age National Scientific Reserve (fig. 3.9). When the melting ice front migrated farther north and east, drainage from the ancestral Lake Superior found a lower eastern outlet, and its outlet through the Bois Brule and St. Croix rivers was abandoned.

Glacial-lake deposits of reddish clay, silt and sand now cover the Lake Superior Lowland on either side of Bayfield Peninsula and, along with glacial till, give most of the area a heavy red clay soil (plates 2 and 4). Near the cities of Superior and Ashland the old lake bed is a plain, but farther inland the surface has a greater slope toward the lake and is deeply eroded. In the plain west of the peninsula, narrow valleys from a few feet to more than 100 feet deep have been cut in the soft,

impervious sediments. The plain east of the peninsula is also dissected but has more sandy ridges and hills.

Superior and Duluth share one of the finest natural harbors on the Great Lakes. It resulted from the drowning of the river mouths at the southwestern tip of Lake Superior, a consequence of the rebound, or rising, of the eastern, or outlet, end of the lake after the melting of the ponderous ice sheet. Wisconsin Point and Minnesota Point, two long, sandy bars built by waves and currents, lie across the mouth of the harbor, protecting it from the large waves of the open lake (fig. 6.38).

East of the harbor the Lake Superior shore presents varied appearances: drowned river mouths and marshy areas; sandstone cliffs, sometimes with wave-carved caves and arches; snug harbors; the Apostle Islands; and stretches of beaches of clean sand or gravel.

Other Glacial Lakes

There were many glacial lakes in Wisconsin—large and small—that were formed when drainage was blocked by the ice sheet (plate 2). Their old lake beds now present surfaces more level than surrounding terrain, and their sediments generally are organically rich and less stony than glacial till (fig. 6.39).

Glacial Lake Wisconsin covered much of the Central Sand Plain. Perhaps it existed more than once, but

166

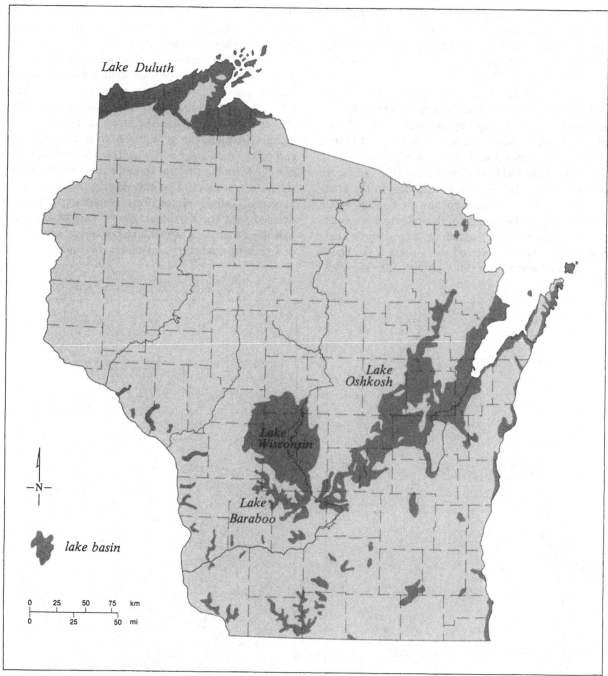

Figure 6.39 Locations of glacial-lake deposits, generalized. Narrow glacial lakes in the Driftless Area were backed-up waters in tributaries of rivers. (Wisconsin Geological and Natural History Survey)

it is known to have formed when the Green Bay Lobe, in its last deep incursion into southcentral Wisconsin, moved across the ancestral Wisconsin River north of the Baraboo Hills and prevented that river from continuing south. The dammed waters became a lake which rose until it was high enough to exit westward to the Mississippi River along what is now the East Fork of the Black River. This broad temporary lake extended over low land from the Green Bay Lobe's terminal moraine and outwash in eastern Adams and Wood counties west to Tomah and the Ironton Escarpment, and

from the Wisconsin Dells area north to Nekoosa (south of Wisconsin Rapids). Its maximum area appears to have been more than 1,800 square miles (4,600 sq km), and its maximum depth about 150 feet.

The lake-bed deposits of Glacial Lake Wisconsin have been partly covered by outwash, wind-blown sand, and alluvium. Many of the higher mounds and buttes now seen on the Central Sand Plain were islands in the glacial lake; lower ones were submerged. The Mill Bluff Unit of the Ice Age National Scientific Reserve in eastern Monroe County contains a number of mounds.

In that area the lake may have been from 60 to 80 feet deep. The former lake level is marked on the sides of some mounds on this plain by erratic boulders that were carried and dropped there by icebergs drifting on the lake.

Glacial Lake Baraboo was a southern bay of Glacial Lake Wisconsin. It held backed-up waters of the Baraboo River, and reached up the river valley to Kendall (southeastern Monroe County) and beyond Hillsboro (eastern Vernon County).

Glacial Lake Oshkosh formed more than once over low land west of the Niagara Escarpment in the area of Green Bay, Lake Winnebago, and the Fox and Wolf rivers. During the last retreat of the ice, while drainage to the northeast was still blocked, the lake made its final appearance, which is referred to as Later Lake Oshkosh. It was about 65 feet (20 m) higher than the level of present Lake Winnebago, which is a remnant of it. Lake Winnebago now is shallow, its greatest depth being only about 20 feet (fig. 6.40).

Many river valleys in Wisconsin contained temporary lakes when the ice sheet dammed them, causing water to pond up into the upper-tributary valleys. These lakes existed for a time and then drained away when the ice dams melted and the lower valleys were open again. One such lake, a relatively small one, was *Glacial Lake Middleton,* which collected in lowlands west of Lake Mendota (central Dane County) when ice from the northeast blocked an east-flowing tributary of the ancestral Yahara River. This temporary lake then drained west toward Black Earth. Part of the Lake Michigan shore was former lake bed, as were the shores of Green Bay, when the lake level was higher.

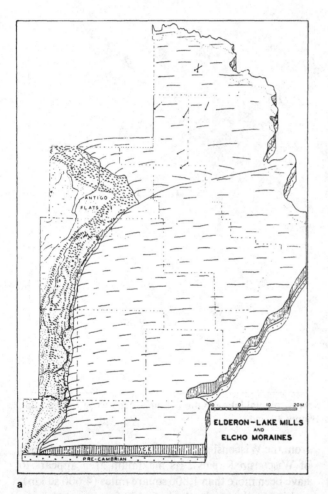 

Figure 6.40 Conditions in part of northeastern Wisconsin, from Lake Winnebago at the southeast corner to Florence County at the north, as depicted by F. T. Thwaites. **(a)** At the time when the Langlade Lobe (north) and Green Bay Lobe (south) were withdrawing, moraines had been left along their retreating edges, and outwash had created the Antigo Flats in Langlade County. **(b)** At a later time, when the Langlade Lobe had gone and the Green Bay Lobe continued its retreat, more recessional moraines had been formed, and Glacial Lake Oshkosh temporarily existed as drainage was dammed. (F. T. Thwaites, 1943)

The Postglacial Terrain and Its Drainage

As the ice pulled back from its moraines and disappeared from the Wisconsin scene an entirely new terrain was revealed. Valleys had been at least partly filled with drift and high features had been scoured lower, making a more even landscape overall. On the other hand, new irregularities were present, the result of uneven deposition of drift. Till plains on which the ice had moved and lain displayed gently undulating swells and swales. Mature pre-glacial drainage patterns had been disrupted, and new drainage was circuitous and youthful. Streams had not had time to erode and straighten their courses. They wandered haphazardly, just flowing to the next lower place, in whatever direction that was. In low spots water collected as ponds and lakes until it was high enough to overflow and then spilled to wherever the lowest outlet led. Rapids, cascades, falls, and winding streams were common. So were marshes and swamps in poorly drained locations.

Over long periods of time, streams grade their courses to remove rapids, marshes, lakes, and circuitous detours. Postglacial streams have accomplished only some of this job. Drainage in areas of new drift is still in a disorganized condition, and that adds to the natural beauty and interest of those regions (fig. 6.41).

Lakes are common throughout the glaciated part of Wisconsin—the result of uneven deposition of drift, disorganized or blocked drainage, and depressions formed by delayed ice-block melting. The area of Vilas, northern Oneida, and eastern Iron counties has an exceptionally large number of small lakes close together. Only a few other well-populated regions of the world have so much of their area in lakes. Another region in Wisconsin having a dense concentration of lakes is the northwest, in the "upper face" part of the Indian Head Country.

Some of the cutting of the Wisconsin River's postglacial channel was done during the draining of Glacial Lake Wisconsin, when that large body of stored water escaped to the southeast as the ice melted back. For some time floods of meltwater continued to augment the river's power. Across the Central Sand Plain the river's new route lay west of its preglacial channel. At the junction of Juneau, Adams, Sauk, and Columbia counties the river flowed over an area of Cambrian sandstone capped by the resistant Ironton Formation. In crossing that area it shifted course, moving from one channel to another, until finally its cutting was concentrated in one channel. There it carved a narrow gorge, the *Wisconsin Dells* (fig. 6.42). Boat trips through the Dells were one of the first major tourist attractions in Wisconsin. In that canyon one can see layering and cross-bedding of Cambrian sandstone, deep weathering of vertical joints, wooded ravines, and many impressive erosional features such as Stand Rock, Coldwater Canyon, Artist's Glen, Witches' Gulch, the Jaws, and the Navy Yard.

The Dells are a little more than 7 curving miles (11 km) long. At the Narrows the gorge's width is only about 50 feet (15 m), but the average width is 150 to 200 feet. The city of Wisconsin Dells, formerly called Kilbourn, separates the Upper Dells to the north from the Lower Dells to the south, and is the focus of one of the state's busiest recreational areas. The Upper Dells are longer and have higher walls than the Lower Dells. The gorge's maximum depth is about 120 feet—about 80 feet being visible wall height and about 40 feet being water depth. The wall height of the Upper Dells was greater before a dam was built at the city of Wisconsin Dells, raising the river level behind it (fig. 6.43).

How the Lower Wisconsin River was diverted around the east end of the Baraboo Hills has been described earlier.

Figure 6.41 Lakes south of Rhinelander, Oneida County. Disorganized drainage resulting from glaciation has created interesting terrain and numerous scenic sites for recreation. (Carl Guell)

Figure 6.42 Part of the Upper Dells of the Wisconsin River north of the city of Wisconsin Dells, looking north. (Carl Guell)

a   b

Figure 6.43 (a) Chimney rock in the Upper Wisconsin Dells as it appeared before the dam was built downstream. (In upper part of picture, at the left side.) Seen from the river, it rose high like a chimney. Notice the cross-bedding and the river's undercutting of the water-worn rock. **(b)** The same location after the dam was built. Water level is higher and Chimney Rock appears less prominent, being almost lost against the gorge wall. High Rock at the right seems not nearly as high either. (H. H. Bennett Studio)

The Yahara River Valley in central Dane County was modified by glaciation too. It now contains the string of lakes historically known as the Four Lakes—from north to south, lakes Mendota, Monona, Waubesa, and Kegonsa. Drift partly filled the preglacial valley; and glacial scouring, drift dams, and stranded ice blocks created the lake basins. Smaller, shallow Lake Wingra in southwestern Madison is another of this family of lakes. The capitol district of central Madison is situated on an isthmus of glacial drift between lakes Mendota and Monona.

Effects of Glaciation in the Central Sand Plain

The bedrock of the Central Sand Plain is Cambrian sandstone, but most of this plain (all but its western part—eastern Jackson and northcentral Monroe counties) is surfaced with glacial deposits.[3] Its center is the lake bed of Glacial Lake Wisconsin, and its eastern part is covered with outwash that was car-ried west from the terminal (Johnstown) moraine of the Green Bay Lobe (plate 2).

An impressive change in terrain from higher rough land to the flat plain is seen in western Waushara County as one travels west or northwest from the curving moraine to the sand plain in the Hancock area. Crossing the moraine hills deposited at the glacier's edge, one looks down upon the low plain stretching out ahead. It is flat because it is surfaced by the sheet of outwash that was spread in front of the melting ice, and, farther beyond, because it is the floor of the old glacial lake. Views of this change can be seen driving north on U.S. Highway 51 from Coloma to Hancock. Going west on State Highway 21 toward Coloma, crossing the crest of the Johnstown terminal moraine, one sees not only the lower sand plain but Friendship Mound and smaller Roche à Cri in the distance.

3. Discussion of the Central Sand Plain in chapter 5 dealt mainly with its being a part of the Cambrian sandstone lowland. Comments about its total character were deferred until its glacial background had been given, for most of it was altered by glaciation.

In the Central Sand Plain the sandstone bedrock and overlying glacial sands and gravels work for good drainage, but because of the land's flatness and slow runoff there are many marshes and swamps. Remains of vegetation that collected in the wetlands over thousands of years have resulted in sizable areas of peat and muck soil. Peat, the first stage in nature's coal-making process, has been used as a low-grade fuel when better fuels were not available, but its main use today is for soil-conditioning. The muck soils, nearly black in color, appear to be exceptionally fertile, but they are not. Some of this land was drained and cultivated in years past, but most did not produce well at that time, and was allowed to revert to its natural state. Recently, however, extensive areas of wet land again have been drained and brought into production successfully. Though these wetlands lack some nutrients, they do have the advantages of being level and easily cultivated, of responding to fertilization, and of being dark in color, which makes them good absorbers of solar heat.

Much of the Central Sand Plain's better-drained land has sandy loam soils, which, by contrast, are light in color and more porous than wetland soils. Since water drains through them quickly they have a dry character, and irrigation is usually required for commercial agriculture. But with the water table close below the surface, water is easy to obtain. To retard the blowing of loose, sandy soil, windbreaks of coniferous trees were planted (fig. 6.44).

In past years the Central Sand Plain's wet and sandy soils appeared to be unsuitable for ordinary field and vegetable crops without costly improvements, so this thinly populated area had the reputation of being one of the state's less productive regions. But the situation is changing. More and more of the wetlands are being drained, and sandy areas irrigated, for it has been learned that when needed fertilizers are added to these soils they are excellent producers of various vegetable and specialty crops.

Even before modern farming techniques were applied in the Central Sand Plain, certain other enterprises thrived under natural conditions. Wild cranberries are native to this area, and commercial varieties are raised with outstanding success. The low, marshy plains provide the environment cranberries prefer—cool summers, and acid soils and water. The ease of acquiring large amounts of water, and of distributing it over wide flat areas, is a boon to production. If frost threatens during the growing season the plants are sprinkled, or their diked bogs are flooded, to protect the berries. Flooding also facilitates harvesting because the berries float. With these ideal conditions, Wisconsin is a leading producer of high-quality cranberries. About three-fourths of its crop comes from the Central Sand Plain. The rest comes from a few localities nearby and from marshy areas in the northwestern and northcentral parts of the state (fig. 6.45).

Figure 6.44 Circular irrigation systems at the University of Wisconsin's Agricultural Experiment Station at Hancock, western Waushara County, in the Central Sand Plain. A well is at the center of an irrigated plot, and pipes that spray water pivot around the plot. Rows of trees were planted as windbreaks. (B-Wolfgang Hoffman, Department of Agricultural Journalism, University of Wisconsin-Madison)

Figure 6.45 Diked cranberry bogs near Warrens, northeastern Monroe County, in the Central Sand Plain. Marshes nearby are natural water reservoirs. Other unused land is in trees. (Carl Guell)

Sphagnum moss requires abundant groundwater and grows naturally in much of the wet land of the Central Sand Plain. Large quantities are harvested from marshes in the western part of the plain. This remarkable plant can absorb many times its weight in water, retains its water-holding ability long after it has been cut, and combats fungi. Sphagnum moss is used to protect plants during shipment, and to wrap grafts

171

Figure 6.46 Profile of an egg-shaped drumlin. Its shape indicates the glacier came from the right (the steeper side) and moved toward the left. (U.S. Geological Survey)

on plants. In shredded form it is used to improve heavy soil, and as a medium in which to start plants.

The Central Sand Plain also provides lumber and other forest products. It has been found to be good land for pine plantations, Christmas tree farms, and cattle raising; and its sands, alluvial gravels, and rocks are used commercially. Tourism is increasing.

This region—with its sandy soils, barren and blowing or covered with brush and woods of oak, jack pine, birch, and aspen; with its bogs and its swamps of tamarack and black spruce—has proved to be not as poor as first believed.

Because most of this large plain remained essentially undeveloped until recently, it has been prized and widely known as a home for wildlife, including deer, smaller animals, and rare birds. Its so-called "wasteland" areas are beautiful in their own way. With expansion of large-scale agriculture and draining of wetlands much of the wildlife habitat will be lost, and gone will be the natural quality of some lands here that were relatively undisturbed through postglacial time.

Figure 6.47 After depositing layers of till the ice sheet pushed them up into folds. These contortions were within a drumlin that was being dug away for its sand and gravel, northeast of the junction of Highway 26 and Interstate 94 in Jefferson County. Not all drumlins contain contortions such as this. The manner in which drumlins formed is not satisfactorily understood. (Ray Pfleger and Dan Zielinski)

Drumlins

Drumlins are smooth, streamlined, oval or elongated hills composed largely of unsorted till, that have their long axis parallel to the direction of the glacier's movement. They formed at the base of the ice sheet and are found on till plains, but only in some areas, and usually in clusters. The lower land between drumlins may be swampy, or swell-and-swale ground moraine. What caused the ice sheet to create drumlins, and only in certain places, is one of glaciation's many unan-

swered questions. The till in drumlins may be clayey, gravelly or sandy. In some drumlins layers of till have been bent and contorted, showing that original deposits were squeezed or torn up by subsequent ice motion. Some drumlins have bedrock cores (figs. 6.46, 6.47).

Wisconsin has one of the world's most outstanding drumlin fields. It is mainly in Dodge, eastern Fond du Lac, Jefferson, Dane, and eastern Columbia counties, in the region that was covered by the southern part of the Green Bay Lobe (fig. 6.48). (The Campbellsport

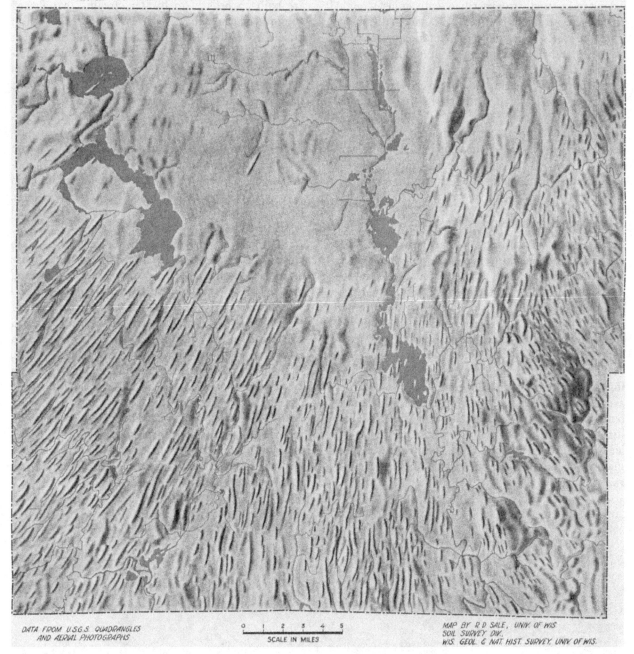

DATA FROM U.S.G.S. QUADRANGLES
AND AERIAL PHOTOGRAPHS

SCALE IN MILES
0 1 2 3 4 5

MAP BY R.D. SALE, UNIV. OF WIS.
SOIL SURVEY DIV.
WIS. GEOL. & NAT. HIST. SURVEY, UNIV. OF WIS.

Figure 6.48 Drumlins are a conspicuous feature in Dodge County, and indicate the direction of ice movement. (Randall D. Sale)

Drumlin Unit of the Ice Age National Scientific Reserve in southeastern Fond du Lac County has fine examples of drumlins [fig. 6.1].) There are drumlins in other parts of the state also.

Because the dimensions of drumlins vary considerably, these hills have been described in a number of ways. A drumlin may be said to look like half of an egg cut lengthwise and laid flat side down; or like the overturned bowl of a spoon; or like a fish; or longer, like a cigar. Many typical drumlins are about 70 feet (20 m)

high, a half mile long, and a quarter of a mile wide. But some measure as high as 140 feet, and others are only several feet high. Their lengths range from tens of feet to two miles. Often two or more drumlins are joined in a composite form.

What is especially significant about drumlins is that they are elongated in the direction of ice movement. An observer looking down upon a drumlin field from the air can see how those hills are aligned parallel to the ice's path. This alignment of drumlins is so char-

173

acteristic that the direction in which the ice moved can be ascertained by it. (It should be noted that the ice did not move strictly from north to south, or just straight along the central axis of a lobe. It fanned outward toward the lobe's sides, and the orientation of drumlins reflects that divergence.) In many drumlins, especially the shorter, egg-shaped ones, the end that faces the direction from which the ice came is typically steeper than the opposite end, which gently tapers away in the direction the ice went.

Drumlins afford good, well-drained cropland, especially on gentle south-facing slopes. Those sides generally have been the first to be cultivated, whereas steeper north-facing sides often remain tree-covered.

Depositional Features of Meltwater

The work of meltwater was as important as that of the ice itself in creating surface features and styling the character of the land. Where the ice sheet had stood still and melted down without moving again, many intricate features built by its meltwater were left intact. In addition to the kames and ice-walled lake plains already described, these include crevasse fillings and eskers.

Crevasse fillings are linear ridges with fairly level tops, generally less than a quarter of a mile long. They are composed of sands and gravels that collected at the bottom of crevasses into which water spilled, depositing cast-off material from the wasting ice. After the ice was gone the ridges of water-deposited material remained (fig. 6.49). Sometimes till also is present in crevasse fillings.

Eskers are long, narrow, steep-sided snakelike ridges shaped in ice tunnels (figs. 6.50, 6.51, 6.52). They are built-up beds of meltwater streams that flowed within or under stagnant ice. Their height in Wisconsin ranges from a few feet to 40 feet or more. Some that are short and discontinuous may resemble crevasse fillings, but others wind along for several miles. Eskers may ascend rather steep grades if the streams that made them were so confined within the ice that they flowed up hill under pressure. Eskers are composed of irregularly stratified sands and gravels, sometimes containing materials as coarse as boulders. Their crests are usually knobby or undulatory. They are found in morainal belts as well as on till plains. A fine example in Wisconsin is Parnell Esker, more than three miles long, in the Kettle Moraine Unit of the Ice Age National Scientific Reserve.

In some cases it is difficult to tell whether a certain landform is an esker, or crevasse filling, or other linear drainage deposit left at the base of a crack in a wasting glacier. Therefore, the more general term *ice-channel filling* is often favored; it is used to mean any ridge of water-deposited materials that developed along a stream bed in or under a stationary glacier (like an esker), or in an open crack between walls of a glacier.

Outwash also played a leading role in fashioning Wisconsin's landscape. This is not surprising when one considers the tremendous amount of water the massive continental ice sheet stored, and how torrential and rampant its outpourings must have been when it was melting. Then there were deluges such as no one alive has ever witnessed, and there also were quietly flowing streams and trickles from the ice.

Figure 6.49 The ridge in the background is a crevasse filling, a linear deposit built along an opening in the ice sheet. Northern Chippewa County. (Robert F. Black)

Figure 6.50 The mouth of a tunnel at the base of a glacier, which is a channel for meltwater flowing from the melting ice. Material deposited in the stream bed may form an esker. (Louie Kirk)

Figure 6.52 A footpath follows the crest of Parnell Esker, in western Sheboygan County near Dundee. This long gravel ridge varies in height from 5 to 35 feet. (Wisconsin Department of Natural Resources)

Figure 6.51 Jersey Esker south of Dundee in the northern unit of Kettle Moraine State Forest. Dundee Mountain, a large kame, is on the horizon. (George Knudsen, Wisconsin Department of Natural Resources)

Figure 6.53 Looking east along State Highway 21 over the level outwash plain that lies in front of the Green Bay Lobe's terminal moraine. The moraine stretches along the horizon. Meltwater coming from the glacier spread this outwash. Near Coloma, Waushara County. (Gwen Schultz)

Boulders and other coarse materials that were too heavy for meltwater to transport were left behind in moraines. In front of the moraines gushed sheets of water, and many diverging or crisscrossing braided streams depositing sands and gravels and carrying away finer material. Much of the outwash was dropped quickly and spread in delta-like fans in front of the mo-

raines. Where many fans merged, broad *outwash plains* were built, some of which are miles across (figs. 6.53, 6.54, 6.55).

Where blocks of dead ice were left standing in front of recessional moraines, outwash collected around and over them. When they melted, hollows pocked the outwash plain. Outwash having these depressions is termed *pitted outwash* (figs. 6.56, 6.57). Occasionally outwash is so pitted that it resembles a rough moraine. Some depressions are fully drained, and some hold ponds, marshes, or lakes. Many of Wisconsin's lakes lie in ice-block depressions in pitted outwash. Many others occupy kettles in moraines, or are formed by moraine dams and by uneven deposition of drift, as noted before.

Figure 6.54 An outwash plain about 5 miles east of Wisconsin Rapids. (Ralph V. Boyer)

Figure 6.55 Gravel is being dug from outwash north of Ladysmith, Rusk County. (Gwen Schultz)

Plate 2 shows the distribution of major areas of outwash in Wisconsin. Outwash covers large areas of northern Wisconsin, where in many places it is more than 100 feet (30 m) thick. There are also thick deposits of sand and gravel outwash buried under much of the till.

Extending southwest from northern Bayfield County for more than 100 miles—across Douglas County and into Washburn County—is a sandy belt 5 to 15 miles (24 km) wide, known as the *Barrens* or the *Pine Barrens*. Much of it is pitted outwash, and used to be barren land with scattered pine trees and dunes. Commercial plantings of pines do well there. Sandy drift is abundant because of the sandstone bedrock of the Lake Superior Lowland over which the ice passed.

Figure 6.56 Pitted outwash near Polar, Langlade County, east of Antigo, about 1927. (F. T. Thwaites, Wisconsin Geological and Natural History Survey)

Figure 6.57 Pitted outwash south of Rosholt in northeastern Portage County. (B-Wolfgang Hoffman, Department of Agricultural Journalism, University of Wisconsin-Madison)

Water drains through the sand quickly, so dryness characterizes the Barrens. However, an underlying nonporous base does permit water to remain in many depressions.

In western Langlade County, where the Langlade and Green Bay lobes abutted against each other and where outwash from both intermingled, are the *Antigo Flats* (fig. 6.40). That broad, sloping plain is one of the better-developed agricultural regions of northern Wisconsin. Its counterpart to the south is *Rock Prairie,* the plain around Janesville and Beloit in Rock County near the end of the interlobate Kettle Moraine, where outwash poured forth (fig. 6.58).

Not all outwash was spread right in front of moraines. Much was carried long distances by rivers draining from the wasting ice sheet. The rivers could not flush away all of the outwash brought into their valleys, especially when the volume of meltwater decreased, so much of the outwash remains on the river valley floors as *valley trains*. The valleys of the lower Wisconsin and Chippewa rivers would be as much as 200 feet (60 m) deeper without their valley trains, and outwash in the Rock River Valley in Rock county is over 300 feet deep (fig. 6.15).

When great floods of meltwater flowed through the main drainageways from the ice sheets, they substan-

a

Figure 6.58 (a) Aerial view of Antigo flats (outwash plain) and the wooded terminal moraine of the Green Bay Lobe, looking northeast. (Joseph J. Jopek) **(b)** Rock Prairie in Rock County beyond the south end of the Kettle Moraine (Robert W. Finley)

b

Figure 6.59 The Mississippi River was a major spillway, carrying meltwater and outwash from the ice sheet. Outwash still clogs its channel with islands and sandbars. Now with less volume and carrying power the river gradually moves sediments of its valley train downstream. This view is south of Prairie du Chien, looking north. (Carl Guell)

tially widened and deepened the bedrock form of the valleys they used. These *spillways* became extra-large valleys, larger than the streams now occupying them could have made handling only today's normal runoff. Spillways commonly were outlets of glacial lakes. (Figs. 6.59, 6.60.)

When the rivers in the spillways were carrying a heavy load of outwash they deposited it on their valley floors; but then at other times, when they were not carrying much outwash, they were able to erode. Then they cut into the outwash already on their floors, leaving relatively level terraces, or benches, of the remaining outwash along their sides. Towns, settlements, highways, and railroads have been built upon those outwash terraces which are well drained and above flood level.

The Mississippi Valley was the main spillway of central United States. During late glacial time the Mississippi River was several times its present size. Into it flowed meltwater from tributaries on both sides, including at times the outflow from Glacial Lake Agassiz. This huge lake impounded by the ice sheet covered northwestern Minnesota, eastern North Dakota, and large parts of Ontario, Manitoba and Saskatchewan. Its discharge into the headwaters of the Mississippi greatly augmented the river's size and erosive power. When the ice sheet had disappeared and Glacial Lake

Agassiz had drained away, the volume and velocity of the Mississippi were much reduced.

Among the Wisconsin spillways tributary to the Mississippi were, from northwest to southeast, the St. Croix, Chippewa, Black, Wisconsin, Sugar, and Rock rivers. Those crossing the Driftless Area left their ribbons of valley-train drift across an otherwise drift-free region (plate 2). The Menominee River carried meltwater to Green Bay.

The navigability of the Wisconsin River is impaired by the shifting sandbars and islands of its valley train. However, in compensation, the river is an inviting down-stream canoe route, and along its sandy banks are many recreation spots (figs. 6.61, 6.62).

The Chippewa River brings its sediments into the Mississippi River Valley, and in postglacial time the Mississippi has not had enough carrying power to keep those sediments moving along. So where the Chippewa enters the Mississippi they have formed a delta which acts as a partial dam. This dam causes a widening of the Mississippi upstream, known as Lake Pepin; the lake is nearly 22 miles long and one to two-and-a-half miles wide. Farther north partial damming of the lower St. Croix River by sediments of the Mississippi has created Lake St. Croix.

The large amounts of glacial outwash left in the spillways will take a long time to reach the sea, if they

Figure 6.60 The confluence of the Wisconsin and Mississippi rivers—looking east up the Wisconsin River, whose valley like that of the Mississippi is filled with outwash. Wyalusing State Park is on the wooded bluff on the right (south) side of the mouth of the Wisconsin River. (Carl Guell)

Figure 6.61 A windblown sandy area along the Wisconsin River near Arena. (Gwen Schultz)

Figure 6.62 Sandbars in the Wisconsin River at Spring Green bridge. The river flows from right to left. (Carl Guell)

ever do, and newly eroded material is continually being added. Rivers now are smaller and move more slowly than in the time of heavy glacial outpourings, but they continue to carry and roll their glacial material gradually downstream.

Loess

During the melting of the ice sheet there were not only great floods, but there were times too when flooding subsided and vast expanses of fresh outwash and bare drift were dry. Then widespread dust storms were common. Winds blew without restraint, and freely picked up dust and deposited it over other areas. Little or no vegetation had returned as yet to slow the wind by friction and to hold down the fine earth materials.

The main sources of dust that settled over Wisconsin were the extensive floodplains of the Mississippi spillway and other floodplains of the river's tributaries. The prevailing wind in these latitudes is from westerly directions. For centuries clouds of windblown silt, called *loess*, were borne over Wisconsin and other downwind areas and deposited there (fig. 6.63). ("Loess" is the Anglicized version of the German word "löss." The origin of this type of wind-deposited silt came to be known first in Europe, where it is widespread.)

Loess is thickest near its source. Hills of the Driftless Area are generously covered with it, especially close to the Mississippi, where it may be up to 60 feet (18 m) thick. It can stand in steep cliffs. The thickness of the loess cover decreases rapidly and irregularly to the east and north, and drops to just a few inches, or is absent, in eastern and northern parts of the state.

Loess is light brown, buff or yellow in color. It contains many different minerals, and forms an exception-ally fertile and friable soil that holds moisture well. Most of Wisconsin's silt loams are loessial soils.

A Comparison of Glaciated and Unglaciated Terrain

Knowing what glaciation has done to Wisconsin's landscape, we cannot help imagining how different it might be if the ice sheets had not come.

The Great Lakes would not exist, and in their place would be river systems draining lowlands. There would be few inland lakes, whereas now Wisconsin has nearly 9,000 lakes of more than 20 acres in size. The major river valleys would not be floored with thick outwash. We can speculate that in regard to topography eastern

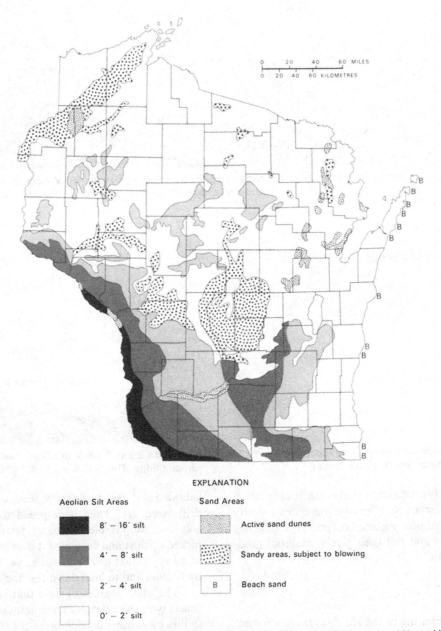

EXPLANATION

Aeolian Silt Areas

■ 8' – 16' silt

▨ 4' – 8' silt

▤ 2' – 4' silt

□ 0' – 2' silt

Sand Areas

▢ Active sand dunes

▨ Sandy areas, subject to blowing

B Beach sand

Figure 6.63 Distribution of aeolian (windblown) deposits in Wisconsin. (Wisconsin Geological and Natural History Survey)

Wisconsin would look much like the Driftless Area does now, since it is a region of similar, slightly tilted sedimentary rocks and of similar climate, and would have experienced comparable weathering and erosion. The Niagara Escarpment, fully revealed, would be a more conspicuous feature than it is, even more prominent than the Ironton Escarpment in the Driftless Area. In fact, without their covering of drift all escarpments in the glaciated Paleozoic section would be more exposed, and they would be more irregular in outline. Had there been no planing by the ice, there would be more outliers; more fragile, bizarre-shaped crags and towers; more buttes and mounds. We would find more caves and sinkholes in the dolomitic areas if glacial erosion and deposition had not taken place.

We can never really know what might have been, but the difference that exists between glaciated and unglaciated terrain now is striking and significant.

In the Driftless Area we see a highly dissected surface, often hilly and rough. Tall bluffs, rugged cliffs, and rocky ledges are common sights, as are jagged crags and isolated mounds. Bedrock outcrops are far more numerous than in drift-covered areas (fig. 6.64). The land is well drained by mature streams tapping all areas, so it is virtually a lakeless region. There lakes and ponds are artificially made. Waterfalls and cascades are gone, because glaciation did not upset stream courses, and because over a long period of time obstacles that once existed in those old channels have been worn away.

In the glaciated area, on the other hand, the topography is smoothed and gently rolling, with shallow valleys; and the glacier-built moraines and hills—not overly steep or angular—blend into the flowing contours. Drainage is irregular. Youthful streams run this way and that around uneven rises of drift, meeting obstructions and drops that create rapids, falls or diversions. Lakes, ponds, marshes, and swamps are familiar features.

Soils (other than loess) in the Driftless Area are residual, developed essentially in place from underlying rock, so they bear a close relationship to that rock. But in the glaciated region the residual soils were removed; a cover of imported, heterogeneous rock material was left instead, and from that the present soils have developed, usually bearing little or no direct relationship to the bedrock (fig. 6.65).

One of the things that makes Wisconsin especially interesting geologically is this side-by-side contrast it offers of heavily glaciated and nonglaciated terrain. Although there are other drift-free "islands" elsewhere, Wisconsin's Driftless Area is particularly impressive because of its large size and its location in a highly developed region. Also, the ice lobes that bypassed it or encroached upon it on all sides were powerful ones. Yet it was never completely surrounded by ice at any one time, or not for long. If it had been it would have lake-bed sediments throughout, which it does not. The known limits of the Driftless Area are not everywhere as precise as they appear on maps, because maps have to be generalized, and because along old-drift borders one cannot clearly discern anymore exactly how far the ice came.

Many a person has come to the Driftless Area to search for evidence that it was glaciated. Some looked

Figure 6.64 In the Driftless Area rock outcrops are common and there is no covering of drift, as in this scene in Grant County. The outcrops are St. Peter Sandstone. (Roger Peters)

Figure 6.65 Major soil regions of Wisconsin. (Wisconsin Geological and Natural History Survey)

but did not find; some thought for a while that they had found "proof," only to have it disproved; some ultimately changed their minds and gave up, or failed to convince others of their "evidence." The Driftless Area has been closely combed for signs of glaciation, and will continue to be, for it presents a challenge. Despite attempts to prove otherwise, this driftless "island" still retains its claim to fame.

Sequel to the Ice Age

Between the Ice Age and postglacial (Recent, or Holocene) time there was a period of transition. That was when herds of long-tusked woolly mammoths roamed the tundra, and the shorter-tusked mastodons browsed through the coniferous forests (figs. 1.27, 1.28). Bones, teeth and tusks of these extinct creatures have been unearthed in many places in Wisconsin, as have remains of other Late Ice Age animals. Swamps in the glaciated region, and rock crevices in the southwestern region were some of the repositories of these fossils (fig. 6.66).

As the climate warmed, tundra gave way to spruce forest; next the pines and hardwood trees came, and, in the southern part of the state, prairie grassland as well.

With the passage of time, soils developed on the once-bare drift surface. The land became better drained as streams graded their courses and worked more efficiently. Many marshes became filled by material washing in from surrounding land and by the accumulation of dead remains of plants that had grown in them. Many lakes shallowed and shrank to become marshes by the same process, and by the encroachment of vegetation growing in from their shores (figs. 6.67, 6.68).

This natural filling-in of marshes and lakes continues today. People are hastening their shrinkage and disappearance, purposely by draining wetlands and by landfill operations, and inadvertently by farming and construction activities which loosen earth materials and speed soil erosion, and so increase deposition in these low places. Nutrients that are added to lakes by runoff from the land or by waste disposal increase aquatic plant growth, which adds to the lake-bottom deposits.

Most of today's lakes were larger or deeper in early postglacial time. Those in northern Wisconsin are younger than those in the southern part of the state, and, in general, have not progressed as far in the lake-dying process. Lakes in southern Wisconsin are younger, in turn, than those farther to the south. In most lakes one can see vegetation growing in from the shore somewhere or taking over shallow places. It is only a matter of time before the lake becomes dry land, either by the filling-in process or by being drained when its outflowing stream wears down its rim. Even so, there are still myriads of marshes and lakes in the glaciated areas of Wisconsin, and they will be plentiful for a long time.

Benefits of Glaciation

The ice did indeed alter the landscape, with significant results that touch us all. It has been said, "We may . . . look at the glaciers as having carried out one of the greatest real estate improvements ever undertaken in Wisconsin."[4]

One way glaciation improved the land was by fashioning a smoother terrain that facilitated settlement, transportation, agriculture, industrial expansion, and general regional development. As we have observed, the ice wore down irregular and rocky features and filled in deep valleys. It created many level areas—of outwash, former glacial-lake beds, and filled-in marshes and lakes. The widespread gently undulating till plains are also readily usable. In the state's glaciated area the percentage of cultivable land is unusually high, and few parts of it are so rough as to discourage agricultural or other commercial activities.

Glaciation added considerably to the fertility and variety of Wisconsin's soil materials. Had the ice not come, most of the soils would be residual only, derived from the underlying rock and none other. Such old soils commonly lack some essential plant nutrients, and those nutrients that the bedrock does supply are released slowly. Weathering of rock takes a long time, and during that time many soluble constituents are leached out and carried away by water draining through the soil. By contrast, drift is a mixture of many kinds of fresh rock, already broken up. When drift was spread over the land it provided soil-making material with a great variety of nutrients. Much of this material is already in the finest particle size, and more has been ground up into small pieces, so weathering can proceed rapidly. Fertility is continuously restored to the soil as the drift's assorted small rocks keep releasing fresh nutrients. Even soils of the Driftless Area, enriched by wind-deposited loess, were improved indirectly by glaciation.

The Swiss naturalist Louis Agassiz (for whom Glacial Lake Agassiz was named) had a strong interest in glaciation which took him to many glaciated parts of the world, including Wisconsin. He called the ice "God's great plough." In 1876 in his *Geological*

4. W. O. Hotchkiss and E. F. Bean, *A Brief Outline of the Geology, Physical Geography, Geography and Industries of Wisconsin,* Wisconsin Geological and Natural History Survey Bulletin no. 67 (Madison: State of Wisconsin, 1925), p. 12.

a

b

c

Figure 6.66 (a) Overview of the site of a mastodon dig. After the glacier withdrew there was a small lake in the low area extending from the center of the picture to the right past the isolated tree. Animals came there for water. The lake gradually became a swamp and filled with marl and peat. The bones of three mastodons have been found where the lake shore once was. Two were found at the left side of the lake and one to the right of the tree. Northwest of Deerfield, eastern Dane County. **(b)** The mastodon dig in progress. **(c)** Some of the finds of the mastodon dig. The man's hand is on a skull. (Zoological Museum, University of Wisconsin-Madison. Don Chandler, photographer)

183

Figure 6.67 A lake (dark oval at right) and two former lakes. The one in the center foreground is becoming a marsh; the dark spot in its center is the remaining open water surrounded by vegetation that is closing in upon it. The other former lake is the light area above the open lake; it has already become a marsh. Eventually the open lake will fill in and become a marsh too. Western Waushara County. (Carl Guell)

Figure 6.68 Vegetation slowly grows in around the shores of a lake and over the water. The lake becomes smaller and fills in and finally disappears. Western Florence County. (Robert W. Finley)

Sketches he explained its role in improving soil in this way:

> When the ice vanished from the face of the land, it left it prepared for the hand of the husbandman. The hard surface of the rocks was ground to powder, the elements of the soil were mingled in fair proportions, granite was carried into the lime regions, lime was mingled with the more . . . unproductive granite districts, and a soil was prepared fit for the agricultural uses of man."[5]

Many people who emigrated from glaciated northern Europe and settled in Wisconsin came here because this area resembled their homeland. They were used to the boulders and drift, the lakes and marshes, the topography and soils, and could adapt their agricultural practices and way of life to this similar region.

Wisconsin's terrain and climate encourage dairy farming. The large tracts of cultivable land interspersed with marshy meadows are ideal for that occupation, and a reason why Wisconsin became the nation's leading dairy state.

The glacier-made wetlands have served this region well, ever since the time when Paleo-Indians first arrived. Wetlands provided habitat for fish, birds, and fur-bearing and game animals upon which the livelihood of those and later people depended, and in many of these wetlands grew wild rice, a staple food. Marshes continue to give pleasure to people interested in their distinctive flora, fauna, and wilderness atmosphere. Peat, which has accumulated in many of them, may at times be used as a fuel—locally and on a small scale—as it has been in the past.

Although drift-covered country has many wetlands, and some clayey areas that are not permeable, drift cover generally has good drainage. Water soaks through it easily because of the open spaces among the loose rock materials. Below the surface, sands and gravels store and distribute abundant supplies of water.

As chapter 4 explained, one geologic reason for the exceptionally reliable and high-quality water supply available in most of Wisconsin is the permeable sedimentary rock, particularly sandstone. Another reason is the drift that thickly covers much of the state: absorbing precipitation like a sponge and letting groundwater seep through freely, it prevents too-rapid runoff, works to purify the water, evens stream flow, and keeps wells supplied. Wisconsin's most flood-prone regions are not in the glaciated sections, but in the Driftless Area where the absorbent drift cover is absent and where many valleys are steep-walled and narrow.

Where drift is thick enough it provides adequate water for farms (including irrigation needs) and small communities. A well drilled into a drift cover of average depth can ordinarily supply enough water for a home; and wells drilled into thick valley trains yield enough water for moderate-size cities. Groundwater from drift is less expensive to obtain, and may even be of better quality, than water from local bedrock. The Precambrian areas especially were helped by the drift cover, as most crystalline rocks can store only small amounts of water, and some none at all.

In addition to the great volume of groundwater in its drift, Wisconsin has a multitude of surface reservoirs—its countless glacier-made lakes, marshes, swamps, and ponds. These too delay runoff and help to regulate stream flow.

Lakes are one of the most precious gifts the glacier bestowed upon Wisconsin and similar areas. In 1918 geologist William C. Alden pointed this out in his historic glacial study of southeastern Wisconsin. By that time the southeastern part of the state was a growing tourist and recreation area because of its nearness to population centers, such as Milwaukee and Chicago. Today his remarks would apply to northern Wisconsin as well. He wrote this of the lakes:

> They attract thousands of visitors from neighboring States with money to spend, besides affording beautiful sites for summer homes and for the recreation of thousands of the State's own citizens. Their contribution to the health and happiness of the people is an asset whose value is not easily computed.[6]

Because glaciation produced the Great Lakes, Wisconsin has "ocean" ports on lakes Michigan and Superior and engages directly in international trade. Also, it has an endless supply of fresh water for its shoreline communities, as long as pollution is controlled. Lake Superior is the last of the Great Lakes to remain clean, because it is the largest, it is uppermost in the chain of lakes, and it has had less population and industry around it. Now, however, in limited areas it too is suffering the effects of pollution, which will continue and spread unless checked.

Sand and gravel, generously supplied by the ice sheet, are among the state's leading economic mineral resources. They are needed in great quantities as aggregate in the construction industry, for making concrete, for roads and highways, and for many other uses.

5. Louis Agassiz, *Geological Sketches* (Boston: James R. Osgood and Co., 1876), p. 99.

6. William C. Alden, *The Quaternary Geology of Southeastern Wisconsin,* U.S. Geological Survey, Professional Paper 106, (Washington, D.C.: U.S. Government Printing Office, 1918), p. 39.

The sorting of drift done naturally by glacial meltwater has endowed Wisconsin with numerous widely distributed deposits of uniform-size sands and gravels which require little or no further sorting before use.

The widespread availability of this construction material has kept Wisconsin's highway-building costs relatively low compared to those of regions where heavy sand and gravel have to be hauled long distances. The initial grading done by glaciers—removing obstacles and smoothing the surface—has also assisted road builders. It is understandable why the automobile had its early success in the glaciated Midwest. At the turn of the century, before the era of concrete and blacktop paving, gravel roads provided the best traction, stability and dryness, being far superior to slippery muddy ones. It is understandable too why Wisconsin has always had an outstanding reputation for well-built and well-maintained highways, including county roads. It was the first state to use a numbered highway system with highway markers. And, despite its relatively large size, it was one of the first states to complete its sections of the original Interstate highway network.

Because of the growing demand for sands and gravels, many glacial features such as eskers, kames, crevasse fillings, and drumlins have been wholly or partly dug away. One reason for establishing the Ice Age National Scientific Reserve was to preserve undamaged examples of these irreplaceable glacier-crafted features (figs. 6.69, 6.70, 6.71).

Other favored sources of sand and gravel are outwash plains and valley trains. Portions of moraines composed of water-sorted drift are used too, but generally moraines are not good sources because they contain many boulders and much unsorted material.

Many Pleistocene clay deposits have economic value. Indians used tempered clay for pottery, and later clays were used for manufactured products, especially bricks and tile. It was found that from certain glacial-lake-bed clay in Milwaukee County attractive, durable, cream-colored bricks could be made. Milwaukee used to be called "The Cream City" because so many of its buildings were constructed of the cream-colored bricks.

Wisconsin's moraines were probably one of the last glacial assets to be appreciated. Because of their ruggedness and their bouldery surfaces they were usually left in their natural condition, tree-covered and unused, though some were grazed. Gradually their value has come to be realized too. Moderately hilly and wooded, with interesting kettles and boulders, they

Figure 6.69 An esker near Myra, Washington County, partially dug away. (Milwaukee Public Museum. Photographer, George Gaenslen)

Figure 6.70 The remains of a kame that was dug away. Eastern Shawano County about 1925. (F. T. Thwaites, Wisconsin Geological and Natural History Survey)

Figure 6.71 A kame near Kewaskum, just outside the Kettle Moraine State Forest, in the process of being removed. (George Knudsen, Wisconsin Department of Natural Resources)

Figure 6.72 A hillside with kettle hole in Summit Lake Moraine, eastern Langlade County, used as a ski slope. (Joseph J. Jopek)

make attractive sites for homes, parks, and recreational developments. Many have the desired height and slope for small ski hills (fig. 6.72).

A 1,000-mile-long Ice Age Trail—an all-season footpath—is being built, to cross the state generally following the terminal moraines of the Green Bay, Langlade, Wisconsin Valley, Chippewa, and Superior lobes, from Door County to Interstate Park on the St. Croix River. Volunteers are doing the construction work. The trail, parts of which cross private land, will connect most units of the Ice Age National Scientific Reserve and many parks. Besides being a scenic-recreational asset, it is an important educational resource (figs. 6.73, 6.74).

Erratic boulders have long been the bane of farmers and others who clear land. It was a tremendous job for the first farmers to remove those heavy rocks. Many were rolled, horse-drawn and sledded into low spots and covered over, or blasted apart, or just dragged to ditches, fence lines or hedgerows where they still stand, overgrown with weeds and brush. Those of liftable size were

1. TUSCOBIA STATE TRAIL

2. RUSK COUNTY FOREST

3. CHIPPEWA COUNTY FOREST

4. CHEQUAMEGON
   NATIONAL FOREST

5. LANGLADE COUNTY FOREST

6. MARATHON COUNTY FOREST

7. DEVIL'S LAKE STATE PARK

8. SUGAR RIVER STATE TRAIL

9. KETTLE MORAINE STATE FOREST

10. AHNAPEE STATE TRAIL

Figure 6.73 The Ice Age Trail, when completed, will be a 1,000-mile footpath following terminal moraines of the last ice sheet, in general, but deviating from them in places. (The Ice Age Park and Trail Foundation)

free, sturdy construction material, and were widely used for foundations of houses and barns, and for fireplaces, fences, silos, and other structures. Often entire buildings were constructed of this fieldstone. Some builders used boulders whole and just cemented them together. Others carefully split them apart to reveal their textures, their mottled and veined patterns, and fresh colors—gray, dark red, pink, black, buff, green—and arranged them by size and color to give a pleasing appearance (figs. 6.75, 6.76). Cobble-size erratics of uniform size have been used in some places to pave paths and streets. Now as you travel through glaciated country that has been settled and farmed for some time you have to look hard to see the once-commonplace erratic boulders lying on the ground. Since they have become rare in populated places they are beginning to be coveted as decorative items.

a

b

Figure 6.74 Segments of the Ice Age Trail, which is popular in winter as in summer. Langlade County. (Joseph J. Jopek)

Figure 6.75 The walls of this former blacksmith shop and community center are built of glacial erratics. In Pella, Shawano County. (Gwen Schultz)

What Wisconsin has gained from glaciation, we see, is not all economic, but is esthetic as well. Homey hills of comfortable size with cozy hollows. Unruly water, churned white, rushing over rocky beds, or dashing in spray over precipitous ledges. The primitive, peaceful sounds in reedy marshes and in swamps of tamarack trees. Chains of gemlike lakes strung out like sparkling necklaces carelessly laid down. These are some of the beautiful scenes bequeathed to Wisconsin by the ice sheet.

Wisconsin's geological heritage is far richer because of the ice sheet's visitations. Though the ice departed several thousand years ago, it still affects our lives in countless ways.

Figure 6.76 An old silo and the lower part of a barn, now gone, were made of erratics—near Delafield in Kettle Moraine country where such rocks are plentiful. (Richard W. E. Perrin)

Epilogue

Most of Wisconsin's landscape may appear modest and plain, as was observed at the start of the book. But monotonous or uninspiring? No, not when understood.

It is hoped that this survey of Wisconsin's geology has given those who live or visit here, or who are otherwise interested in Wisconsin, a better appreciation of this state's geologic foundations and their many important effects upon our lives. Wisconsin's physical resources and attractions are still not fully realized. But as this realization spreads, people become more aware of the natural assets here; tourism grows; more recreational and educational nature-study areas are established; and interest in the resource base is renewed.

In time, the geologic story as presented here is sure to be revised and told in greater detail. Future discoveries, more precise mapping, improved research techniques, and a larger number of people engaged in seeking and supplying information will provide new knowledge and new insights.

It seems appropriate to quote something Lawrence Martin wrote for the preface of his 1916 edition of *The Physical Geography of Wisconsin:*

> It will lead to better habits of thought and reasoning if the readers are taken into the confidence of the author and allowed to realize that by no means every question regarding the physical geography of Wisconsin is to be regarded as settled. One set of facts may favor one interpretation, other facts another, but any person may go into the field and find additional phenomena which support either of the suggested explanations, or lead to an entirely new one.[1]

And the "person" he refers to need not be a trained specialist. Nonprofessional, lay observers can make significant contributions, as many have. Geologist Robert F. Black attested to that when he wrote: "It is through tourists and landowners, as well as scientists, that new data on the geologic past come to light every day."[2] All of us can be students and discoverers of Earth's "gold mine" of exciting information.

In concluding, we cannot help thinking back to the beginnings. Here ages ago the once-dominant trilobites and cephalopods crawled, leaving their trails and fossil remains. Now we are making our own trails and burying our own fossils. On beaches we watch waves forming delicate ripple marks just like those that formed a billion years ago and were molded in rock. We think how many of those wet sand grains now being arranged by the waves were part of the Paleozoic rocks in the distant past, or of the Precambrian rocks before that; and how more recently many were gripped in glacial ice; and still they are rolled and redeposited as the planet orbits and its waters wash and, in other realms, its glaciers move, on and on.

Where the primeval worms dug their holes, people now tunnel too, and quarry, and aspire to grand accomplishments. Where colonies of tiny organisms built ancient coral reefs composed of myriads of small joined dwelling units, we now similarly build our apartments, houses and cities, even using the very rock made by those long-vanished marine creatures.

If we look at this region, or any region, in terms of its geologic base and the natural processes that made it what it is, we can more intelligently trace the course of its historical and modern development. And as we travel across an area, even over familiar routes, and as we dig in our gardens and plow our fields, or look at old stone buildings, we may see things in a somewhat different light. Perhaps the gravels in the stream, the outcrops in road cuts, the rock rubble beneath a cliff, the shape of a hill or valley, will have new meaning. Perhaps we shall respond more perceptively and wisely to changes in the physical environment taking place before our eyes.

1. Pp. viii–ix; 1932 edition, p. ix.

2. Robert F. Black, *Geology of Ice Age National Scientific Reserve of Wisconsin,* National Park Service Scientific Monograph Series no. 2, (Washington, D.C.: U.S. Government Printing Office, 1974), p. iv.

Glossary

Note: In this book special terms or words that need clarification are defined or explained where introduced or where they make their first main appearance. Therefore, by consulting the index you may locate a place in the book where there is further explanation or application of a glossary term, or where there is explanation of a term that is not in the glossary, perhaps because it appears only once.

agate A variety of fine-grained quartz, the colors of which are arranged in bands (as in Lake Superior agate) or in other designs. Commonly formed in cavities in volcanic rock.

alluvium Unconsolidated material such as clay, silt, sand, or gravel that has been deposited during recent time by running water, as in a stream bed, delta, floodplain, or alluvial fan. Stream-deposited sediment.

amygdule A bubble cavity, or vesicle, in an igneous rock that is filled with secondary minerals such as quartz, calcite or copper. A mineral filling that formed in vesicles of lava flows.

aquifer A permeable water-bearing stratum below the earth's surface that contains enough empty space to hold significant quantities of groundwater, and through which groundwater moves freely. Generally an impervious layer lies above and below the aquifer, confining the groundwater and allowing it to travel long distances.

argillite Compact clayey rock derived from mudstone or shale, usually weakly metamorphosed; intermediate between shale and slate. Differs from shale in being more firmly cemented, and from slate in having no slaty cleavage.

banding Layering in rock, where layers are of different color or appearance.

bedrock Solid rock that underlies soil or other unconsolidated surface material.

boulder train Erratic boulders strewn for some distance in the lee of an outcrop by a glacier that passed over the area. It serves as an indicator of which direction the glacier moved.

brachiopods A class of animals having a bivalve shell that is bilaterally symmetrical.

breccia A coarse-grained rock consisting of angular, broken rock fragments (greater than 2 mm in diameter) cemented together. The consolidated equivalent of rubble. Similar to conglomerate rock except that most of its fragments have unworn corners and sharp edges. It may be of any composition or origin.

calcareous Containing a considerable amount of calcium carbonate.

case-hardened rock A porous rock, usually sandstone, whose surface is coated with a cement that formed by evaporation of mineral-bearing solutions.

continental drift The concept that continents slowly drift, or move, across the earth's surface, changing form and location over long periods of geologic time. See plate tectonics.

coulee In Wisconsin, a narrow, steepsided valley closed at the upper end, having a flat sandy or silty floor which is drained by a small stream that typically flows only in spring or intermittently. Coulees are tributary to a valley of a large permanent stream or to another coulee.

crevasse An open crack or fracture in a glacier, vertical or nearly so.

crevasse filling A linear ridge with fairly level top, composed of stratified sand and gravel that collected at the bottom of a glacier crevasse into which meltwater spilled, depositing sediment.

cross-bedding The arrangement of layers of stratified rock where minor beds are inclined more or less regularly in sloping or concave form, lying at an angle to the main bedding plane or depositional surface. Caused by local changes in directional movement of the depositing current.

cuesta A ridge having a long, gentle slope on one side and a steep slope or escarpment on the other side—the result of differential erosion of gently inclined sedimentary strata. The gentle slope agrees generally with the dip of the resistant bed that forms the ridge.

dalles French plural of *dalle*, meaning "gutter". Dells. A river gorge with nearly vertical walls. A steep-walled part of a stream channel.

differential erosion The more rapid erosion of certain rocks or portions of the earth's surface as compared to the slower erosion of others.

dip The angle or degree of tilt of a rock layer from a horizontal plane.

divide Relatively high land that divides stream flow, so heads of streams flow in opposite directions.

dolomite rock A carbonate sedimentary rock consisting mainly of the mineral dolomite or approximating that mineral in composition; or a kind of limestone or marble rich in magnesium carbonate. Limestone that contains a predominantly large amount of the mineral dolomite. Formerly called "magnesian limestone."

drainage basin The area drained by a given stream.

drift (glacial drift) Rock debris of any composition or mixture, including boulders, gravel, sand, and clay, transported by a glacier and deposited either by the glacier or by meltwater flowing from it.

drumlin An oval or elongated, streamlined hill of glacial till, formed at the base of an ice sheet on a till plain, oriented in the direction the glacier moved, typically with steeper end facing the direction from which the ice came.

end moraine A moraine deposited at the end of a glacier, as distinguished from ground moraine deposited beneath a glacier. Commonly called just "moraine."

epicontinental Lying upon a continent or upon a continental shelf.

erosion The processes whereby rock material is moved, in solid form or in solution, from one place to another. The picking up and carrying away of rock material. The main agents of erosion are water, wind and glaciers.

erratic A rock fragment that has been transported by a glacier and dropped some distance from the outcrop from which it was derived, in a region of different bedrock.

escarpment A long, more-or-less-continuous line of cliffs or relatively steep slopes facing in one general direction. Caused by differential erosion or faulting.

esker A low, narrow ridge, often sinuous, composed of crudely stratified sand and gravel deposited by meltwater in a tunnel within or at the base of a stagnant glacier, and retaining its form after the ice has melted.

extrusion The flowing of molten rock from within the earth onto the earth's surface as lava. Also the rock formed by that process, such as a lava flow.

extrusive rock Igneous rock formed from molten material that cooled after reaching the earth's surface, as opposed in intrusive rock, which cooled below the surface.

fault A line or zone of rock breakage along which displacement has occurred. A fracture along which slipping movement has taken place, so that one section of rock has been displaced relative to the other, in any direction.

folding The bending of layered rock formations caused by compression. The distortion of rocks into wavelike forms.

formation As applied to stratigraphy, the basic unit in local classification of rocks. Contiguous rock strata together having common properties of composition or age or both. A body of rock that has some degree of similarity throughout. Formations are combined into "groups" and subdivided into "members". See plate 5.

As applied to landforms, a formation is a topographic feature significant in itself, or noteworthy for some reason.

fossil Any remnant, imprint or trace of a plant or animal that has been naturally preserved in the earth since some past geologic time. Any evidence of past life, including shells, bones, impressions of vegetation or animals, and tracks and burrows made by animals.

gabbro A coarse-grained, dark igneous rock. The intrusive equivalent of basalt.

glacial lake A lake of glacier meltwater and surface runoff whose existence results from the presence of a glacier or from the scouring or deposition done by a glacier. Commonly, a temporary lake resulting from the damming of a drainage outlet by a glacier; the lake drains away when the glacier recedes or melts.

glacial stage A time during the Ice Age when ice sheets formed and expanded.

glacier A mass of ice, formed from compacted and recrystallized snow, that slowly moves under pressure of its own weight or by downslope creep, or that has so moved in the past.

gneiss A coarse-grained rock, usually metamorphic, typically with streaked appearance. Generally composed of bands which differ in color and composition; bands of mainly granular minerals alternate irregularly with bands of mainly foliated minerals.

graben A block of the earth's crust, generally longer than it is wide, that has dropped along faults on either side. A down-dropped fault block. (German word for "ditch" or "trench.")

graywacke An impure, "dirty" sandstone. Dark, hard, coarse-grained sandstone consisting of poorly sorted fragments of several different minerals and rocks bound together by a fine-grained matrix.

greensand A sand having a greenish color, generally because of the presence of grains of glauconite, or a sandstone consisting of greensand.

ground moraine The rock debris that was carried by a glacier and deposited under the ice as it moved or melted. A broad layer of till whose surface is an undulating plain with low relief and gentle slopes (swell-and-swale topography).

group A major unit of rock stratigraphy next higher in rank than "formation," consisting of two or more associated or contiguous rock formations having features in common. See plate 5.

headwaters Upper tributaries of a stream.

horst A block of the earth's crust, generally longer than it is wide, that has been uplifted, along faults, relative to the rocks on either side. Opposite of a graben.

ice cap A wide area of glacial ice or a complex of merged glaciers, smaller than an ice sheet.

ice-channel filling Any ridge of meltwater-deposited materials that formed along a stream bed in or under a stationary glacier or between walls of such a glacier, such as an esker or crevasse filling.

ice sheet the largest form of glacier. A large sheet-like glacier that moves outward in all directions. Example: Antarctica ice sheet, or that which existed over North America during the Ice Age.

inlier An isolated exposure of bedrock surrounded by rocks of younger age.

interglacial stage A relatively warm time during the Ice Age when glaciers retreated, when ice sheets of the northern hemisphere withdrew or disappeared. A relatively warm time interval between major glacial advances of the Pleistocene Epoch.

interlobate moraine A moraine that formed between two adjacent lobes of a glacier.

intrusion Igneous rock that fills cracks or cavities in older rock, having forced its way into those locations while in a molten or plastic state before solidifying there. The entry of molten rock or magma into or between older rock formations.

intrusive rock Igneous rock that was injected into older rocks as molten material, and solidified there underground.

iron-formation A thin-bedded, low-grade sedimentary iron ore (containing at least 15% iron), usually of Precambrian age, in which layers rich in iron irregularly alternate or mingle with layers rich in chert.

joint A fracture in rock along which no appreciable movement has occurred.

kames Steep-sided hills or mounds of water-sorted sands and gravels that were built where streams of meltwater draining from stagnant glacial ice dropped their load of sediment as their velocity decreased. They may have formed against the side of a wasting glacier or at the bottom of holes in motionless ice. A low hill, often conical, of stratified drift formed in contact with glacial ice.

kettle (kettle hole) A depression in drift (in a moraine or on a plain), typically caused by the delayed melting of a detached mass of glacial ice that had been partly or totally buried in the drift.

lava Molten rock flowing out at the earth's surface.

limonite A brown or yellowish-brown iron ore.

lithology The study of rocks. The physical character of a rock formation.

lobe A rounded protuberance of a glacier. A bulge of ice pushing outward from the main body of a glacier along its margin.

loess (*löss*—German) Wind-transported silt, mainly of Pleistocene age. A sediment—typically unstratified, unconsolidated and calcareous—composed of silt-size particles, deposited mainly by wind; typically found in areas peripheral to, and downwind from, those that were covered by the last ice sheet. Its source was dry areas and fresh outwash plains and other bare glacial deposits subject to wind erosion. Loess generally has a uniform, buff color and lacks visible layering.

magma Hot, molten rock material that originates within the earth and may move upward through the earth's crust toward the surface. When this fluid material cools and solidifies it forms igneous rock.

massive In geology, descriptive of rocks occurring in thick beds, or of homogeneous rocks lacking particular structure.

meltwater Water from melting glacier ice which flows on, in or away from the glacier.

member In rock classification, a division of a "formation," generally of a distinct character or of only local extent. A rock stratigraphic unit of subordinate rank; a part of a formation having special characteristics.

metamorphic rock Rock that has been changed by heat, pressure, movement, and/or solution.

mineral A naturally formed chemical element or compound, usually inorganic, having a definite chemical composition and usually a characteristic crystal form.

moraine Assorted, unconsolidated rock material deposited by a glacier. A landform composed of such rock material deposited by a glacier; generally an irregular band of hills that has the outline of the glacier's margin. See end moraine, ground moraine, interlobate moraine, recessional moraine, terminal moraine.

oolite A small, sand-size, more-or-less spherical grain (0.25 to 2.00 mm in diameter) of concentric, usually calcareous layers around a tiny core which is commonly a quartz sand grain or shell fragment. A sedimentary rock, usually limestone or dolomite, made up chiefly of oolites cemented together.

organic Originated from plants or animals.

orogeny The process of the forming of mountainous belts involving folding and faulting.

outlier A part of a rock formation lying at a distance from the main body to which it was originally joined, the section between having been removed by erosion.

outwash Sand and gravel carried from glaciers and deposited by meltwater streams. The smaller materials washed out from a glacier by its meltwater; usually sorted and stratified.

outwash plain A broad expanse of glacial outwash. A flat or gently sloping sheet of sand and gravel spread by many meltwater streams flowing out from the front of a glacier.

pitted outwash Outwash with a pitted surface. Outwash marked with many irregular depressions caused by isolated blocks of glacier ice that melted following the deposition of the outwash.

plate tectonics The concept that the earth's outer crustal shell—including both the continents and ocean floors—is divided into a number of large rigid plates, or broad blocks, which very slowly move about horizontally, divide, and join with one another, causing volcanic activity, deformation of bedrock, and mountain-building along their edges where they meet.

pothole A hole carved in rock at the bottom of a swift, eddying stream that carries sediment, usually in falls or rapids.

quartz One of the commonest rock-forming minerals, silicon dioxide. It occurs either in colorless or tinted, 6-sided crystals or in crystalline masses. It is the main constituent of sandstone and quartzite, and forms the major proportion of most sands.

quartzite A metamorphic rock consisting mainly or entirely of quartz. Most quartzite is quartz sandstone that has been metamorphosed, thoroughly cemented by silica. Quartzite is very resistant—extremely hard, and highly immune to chemical decay.

recessional moraine A moraine built along the edge of a glacier as it paused during its retreat.

relief The difference in elevation between the high and low points of a land surface. Local relief: The difference in elevation between the highest and lowest points in a given area.

rhyolite A light-colored igneous rock of extrusive (volcanic) origin, chemically equivalent to granite but not as visibly crystalline.

rift A narrow opening in rock made by cracking or splitting.

schist A medium-grained to coarse-grained rock that during metamorphism acquired a foliated structure; it shows a cleavage along which it can easily be split into flakes or slabs. It may have originated from either igneous or sedimentary rock. The mineral mica, which has sheetlike structure, is usually prominent.

sedimentary rocks Rocks formed most commonly by the layered accumulation—usually in water—of transported rock debris (sandstone, shale, conglomerate), and of chemically precipitated material (limestone), and of the remains of animals and plants (certain limestones, and coal); and by other less common processes.

series A term used in the classification of stratigraphic rock units based upon time, next in rank below "system". A time-stratigraphic subdivision of a major "system" of rocks. See plate 5.

shield A large area of mainly Precambrian igneous and metamorphic rock that has been relatively stable for a long time. It surface, having been subjected to long erosion, is generally smooth, and is commonly convex like a battle shield. An ancient core of a continent, a nucleus around which younger rocks have formed.

silica Silicon dioxide. A compound of oxygen and silicon that is resistant to change.

siltstone A fine-grained sedimentary rock consisting mainly of grains of silt size—from 1/256 mm to 1/16 mm in diameter. Intermediate between sandstone, which has larger grains, and shale, which has finer ones.

spillway (glacial) A river valley that was a main drainageway for meltwater flowing from a glacier, now commonly much larger than the valley the present river would have carved.

stratum (plural strata) A bed, or layer, of one kind of sedimentary rock.

stream In geology, any body of running water, of any size from small rill to large river, moving under gravity flow in a natural channel.

substage A time of pronounced glacial advance or retreat within a stage.

swales the shallow depressions between swells on an undulating till plain, often marshy.

swells The low hills with gentle slope on a till plain, separated by swales.

syncline A downfold or trough in which rock strata dip inward from both sides toward the axis.

system A major time unit, applicable on a worldwide basis, used in the chronological classification of rocks of Paleozoic and younger age. A term that designates rocks formed during a major chronologic period, as, for example, the Ordovician System. A system is subdivided into smaller time units called "series." See plate 5.

taconite A bedded, chert-like rock of the Lake Superior region, which is high enough in iron content (about 25%) to constitute a low-grade iron ore. This ore must be artificially enriched to be economically usable. The rock is finely crushed; iron-rich minerals are separated by a magnetic or other process, and concentrated into more easily shipped pellets of higher (about 60%) iron content.

talus An accumulated heap or sloping sheet of loose, coarse rock waste at the base of a cliff or steep slope from which it has fallen.

terminal moraine A moraine built along the edge of a glacier at its most advanced position, or built along lines marking the termination of an important glacial advance.

till Material carried and deposited directly by a glacier (not by its meltwater). The unstratified, unsorted part of glacial drift, usually a heterogeneous mixture.

till plain An extensive area having a flat to undulating surface composed of till. A plain with a surface of till.

tuff Consolidated volcanic ash composed mainly of fragments (less than 4 mm in diameter) produced directly by volcanic eruption.

unconformity A surface of erosion or nondeposition—usually erosion—that separates younger strata from older rocks. A gap in the depositional sequence of sedimentary rocks, or a break between the surface of eroded igneous rocks and overlying younger sedimentary strata.

unit In stratigraphy, a general term for a rock layer that formed under uniform conditions and is individually identifiable.

valley train A long body of outwash deposited along a valley floor by a meltwater stream that drained away from a glacier. A long, narrow body of outwash confined within a valley.

vesicle A small, enclosed space, or cavity, in igneous rock, formed by the expansion of a bubble of gas or steam during the solidification of the rock. A cavity in a lava flow formed by the entrapment of a gas bubble during solidification of the lava. The pore left in a lava flow by a gas bubble that formed in the molten rock.

warping A gentle bending of the rocks of the earth's crust without pronounced folding, on a regional scale.

weathering The processes that cause rocks at or near the earth's surface to chemically decay or mechanically disintegrate. Chemical action, including that of water, atmospheric gases, and plant and animal materials, cause rock to decompose. Physical forces, such as those resulting from temperature changes and plant growth, cause rock to break up into smaller fragments.

Selected References

Exhaustive lists of references to geologic publications, largely technical and specialized, can be found elsewhere, even in publications given below. Here, in this limited selection for the average reader, the references are mainly, but not exclusively, ones that will be of interest to the lay public, that pertain to certain material in this text, and that are readily accessible. Many are in local libraries, and many can be obtained from the Wisconsin Geological and Natural History Survey at the University of Wisconsin in Madison.

General (Recent and Historical)

Chamberlin, T. C. and others, 1877–1883. GEOLOGY OF WISCONSIN: SURVEY OF 1873–1879. 4 vol. Madison: Commissioners of Public Printing.

Crowns, Byron, 1976. WISCONSIN THROUGH 5 BILLION YEARS OF CHANGE. Wisconsin Rapids: Wisconsin Earth Science Center.

Dutton, Carl E., ed., 1976. MINERAL AND WATER RESOURCES OF WISCONSIN. Washington: U.S. Government Printing Office.

Hall, James and Whitney, J. D., 1862. REPORT ON THE GEOLOGICAL SURVEY OF THE STATE OF WISCONSIN. Printed by the authority of the legislature of Wisconsin.

Hotchkiss, W. O. and Bean, E. F., 1925. A BRIEF OUTLINE OF THE GEOLOGY, PHYSICAL GEOGRAPHY, GEOGRAPHY, AND INDUSTRIES OF WISCONSIN. Bull. 67. Wis. Geological and Natural History Survey.

Martin, Lawrence, 1916. THE PHYSICAL GEOGRAPHY OF WISCONSIN. 2nd ed., 1935. Bull. 36. Wisconsin Geological and Natural History Survey. Paperback reprint, 1965, University of Wisconsin Press.

Paull, Rachel Krebs and Paull, Richard A., 1977. GEOLOGY OF WISCONSIN AND UPPER MICHIGAN, INCLUDING PARTS OF ADJACENT STATES. Dubuque, Iowa: Kendall/Hunt.

Paull, Rachel E. and Paull, Richard A., 1980. WISCONSIN AND UPPER MICHIGAN, INCLUDING PARTS OF ADJACENT STATES: HIGHWAY GUIDE. Dubuque, Iowa: Kendall/Hunt.

GEOSCIENCE WISCONSIN, a journal published by the Wisconsin Geological and Natural History Survey, University of Wisconsin, Madison, Wisconsin.

References Sources

Reinhard, Christine, 1982. BIBLIOGRAPHY AND INDEX OF WISCONSIN GEOLOGY, 1698–1977. Madison: Wisconsin Geological and Natural History Survey.

Current LIST OF PUBLICATIONS, Wisconsin Geological and Natural History Survey, University of Wisconsin, Madison, Wisconsin.

List of reports and maps of the U.S. Geological Survey, U.S. Government Printing Office, Washington, D.C.

Precambrian

Aldrich, H. R., 1929. THE GEOLOGY OF THE GOGEBIC IRON RANGE OF WISCONSIN. Bull. 71. Wis. Geological and Natural History Survey.

Craddock, Campbell, 1972. "Regional Geologic Setting" in GEOLOGY OF MINNESOTA: A CENTENNIAL VOLUME. P. K. Sims and G. B. Morey, eds. St. Paul: Minnesota Geological Survey.

Craddock, Campbell, Mooney, H. M., and Kolehmainen, Victoria, 1970. SIMPLE BOUGUER GRAVITY MAP OF MINNESOTA AND NORTHWESTERN WISCONSIN. Misc. Map M–10. Minnesota Geological Survey.

Dalziel, I. W. D., and Dott, R. H., Jr., 1970. GEOLOGY OF THE BARABOO DISTRICT, WISCONSIN: A DESCRIPTION AND FIELD GUIDE INCORPORATING STRUCTURAL ANALYSIS OF THE PRECAMBRIAN ROCKS AND SEDIMENTOLOGIC STUDIES OF THE PALEOZOIC STRATA. Information Circular no. 14. Wis. Geological and Natural History Survey.

Dott, R. H., Jr., and Dalziel, W. D., 1972. "Age and Correlation of the Precambrian Baraboo Quartzite of Wisconsin." JOURNAL OF GEOLOGY, vol. 80, no. 5.

Dutton, Carl E., 1971. GEOLOGY OF THE FLORENCE AREA, WISCONSIN AND MICHIGAN. U.S. Geological Survey Prof. Paper 633. U.S. Government Printing Office, Washington, D.C.

Dutton, Carl E. and Bradley, Reta E., 1970. LITHOLOGIC, GEOPHYSICAL, AND MINERAL COMMODITY MAPS OF PRECAMBRIAN ROCKS IN WISCONSIN (with 15 p. text). Misc. Geol. Investigations, Map I–631. U.S. Geological Survey, Washington, D.C.

Green, J. H., 1968. THE TROY VALLEY OF SOUTHEASTERN WISCONSIN. Prof. Paper 600-C. U.S. Geological Survey, Washington, D.C.

Greenberg, Jeffrey K. and Brown, Bruce A., 1983. LOWER PROTEROZOIC VOLCANIC ROCKS AND THEIR SETTING IN THE SOUTHERN LAKE SUPERIOR DISTRICT. Memoir 160. Geological Society of America.

LaBerge, Gene L. and Myers, Paul E., 1983. PRECAMBRIAN GEOLOGY OF MARATHON COUNTY, WISCONSIN. Information Circular no. 45. Wis. Geological and Natural History Survey.

Leith, C. K., Lund, Richard J., and Leith, Andrew, 1935. PRE-CAMBRIAN ROCKS OF THE LAKE SUPERIOR REGION. Prof. Paper 184, inc. map. U.S. Geological Survey.

Maass, R. S. and Van Schmus, W. R., 1980. PRECAMBRIAN TECTONIC HISTORY OF THE BLACK RIVER VALLEY, field guide. Prepared for Institute on Lake Superior Geology, University of Wisconsin–Eau Claire. Wis. Geological and Natural History Survey.

Mudrey, M. G., Jr., 1979. MIDDLE PRECAMBRIAN GEOLOGY OF NORTHERN WISCONSIN. Field trip guide book no. 4. Wis. Geological and Natural History Survey.

Musolf, Gene E., 1984. "The Crystalline Monadnocks of North-Central Wisconsin." TRANSACTIONS OF THE WISCONSIN ACADEMY OF SCIENCES, ARTS AND LETTERS, vol. 72.

Myers, Paul E. and others, 1980. PRECAMBRIAN GEOLOGY OF THE CHIPPEWA VALLEY, field guide. Prepared for Institute on Lake Superior Geology, University of Wisconsin–Eau Claire. Wis. Geological and Natural History Survey.

Myers, Paul E. and others, 1984. THE WAUSAU SYENITE COMPLEX, CENTRAL WISCONSIN, field trip. Prepared for Institute on Lake Superior Geology. University of Wisconsin–Eau Claire.

Olson, Edwin E., "History of Diamonds in Wisconsin." GEMS AND GEMOLOGY, spring, 1953.

Penman, John T., 1977. "The Old Copper Culture: An Analysis of Old Copper Artifacts." THE WISCONSIN ARCHAEOLOGIST, vol. 58.

Salisbury, R. D. and Atwood, W. W., 1900. THE GEOGRAPHY OF THE REGION ABOUT DEVIL'S LAKE AND THE DALLES OF THE WISCONSIN. Bull. 5. Wis. Geological and Natural History Survey.

Smith, Eugene I., Paull, R. A., and Mudrey, M. G., eds., 1978. PRECAMBRIAN INLIERS IN SOUTH-CENTRAL WISCONSIN. Field trip guide book no. 2. Wis. Geological and Natural History Survey.

Vierthaler, Arthur A. "Wisconsin Diamonds." WISCONSIN ACADEMY REVIEW, spring, 1958.

Vierthaler, Arthur A. "Wisconsin Diamonds." GEMS & GEMOLOGY, fall, 1961.

Wold, Richard J. and Hinze, William J., 1982. GEOLOGY AND TECTONICS OF THE LAKE SUPERIOR BASIN. Memoir 156. Geological Society of America.

Paleozoic

Heyl, A. V., Jr. and others, 1959. THE GEOLOGY OF THE UPPER MISSISSIPPI VALLEY ZINC-LEAD DISTRICT. Prof. Paper 309. U.S. Geological Survey.

Heyl, A. V., Broughton, W. A., and West, W. S., 1978. GEOLOGY OF THE UPPER MISSISSIPPI VALLEY BASE-METAL DISTRICT, 3rd ed. Information Circular no. 16. Wis. Geological and Natural History Survey.

Mather, Cotton and others, 1975. UPPER COULEE COUNTRY. Pierce County Geographical Society. Prescott: Trimbelle Press.

Mather, Cotton. "Coulees and the Coulee Country of Wisconsin" in WISCONSIN ACADEMY REVIEW, vol. 22, Sept., 1976.

Mudrey, M. G., Jr., ed., 1978. UPPER MISSISSIPPI VALLEY BASE-METAL DISTRICT. Field trip guide book no. 1 (companion volume to Information Circular no. 16). Wis. Geological and Natural History Survey.

Nelson, Katherine G., ed., 1977. GEOLOGY OF SOUTHEASTERN WISCONSIN. Guidebook for 41st Annual Tri-state Field Conference. University of Wisconsin–Milwaukee.

Odom, E. Edgar and others, 1978. LITHOSTRATIGRAPHY, PETROLOGY, AND SEDIMENTOLOGY OF LATE CAMBRIAN–EARLY ORDOVICIAN ROCKS NEAR MADISON, WISCONSIN. Field trip guide book no. 3. Wis. Geological and Natural History Survey.

Ostrom, Meredith E., 1967. PALEOZOIC STRATIGRAPHIC NOMENCLATURE FOR WISCONSIN. Information Circular no. 8. Wis. Geological and Natural History Survey.

Ostrom, Meredith E., and Davis, Richard A., Jr., 1970. FIELD TRIP GUIDEBOOK FOR CAMBRIAN–ORDOVICIAN GEOLOGY OF WESTERN WISCONSIN. Information Circular no. 11. Wis. Geological and Natural History Survey.

Thwaites, F. T., 1960. "Evidences of Dissected Erosion Surfaces in the Driftless Area." TRANSACTIONS OF THE WISCONSIN ACADEMY OF SCIENCES, ARTS AND LETTERS, vol. 49.

Trewartha, Glenn T. and Smith, Guy-Harold, 1941. "Surface Configuration of the Driftless Cuestaform Hill Land." ANNALS OF THE ASSN. OF AMERICAN GEOGRAPHERS, vol. 31, no. 1.

Pleistocene and Recent

Alden, William C., 1918. THE QUATERNARY GEOLOGY OF SOUTHEASTERN WISCONSIN with a chapter on the older rock formation. Prof. Paper 106., U.S. Geological Survey.

Black, Robert F., 1969. "Glacial Geology of Northern Kettle Moraine State Forest, Wisconsin." TRANSACTIONS OF THE WISCONSIN ACADEMY OF SCIENCES, ARTS AND LETTERS.

Black, Robert F. and others, 1970. PLEISTOCENE GEOLOGY OF SOUTHERN WISCONSIN, a field trip guide with special papers. Information Circular no. 15. Wis. Geological and Natural History Survey.

Black, Robert R., 1974. GEOLOGY OF ICE AGE NATIONAL SCIENTIFIC RESERVE OF WISCONSIN. National Park Service Scientific Monograph Series no. 2. Washington: U.S. Government Printing Office.

Clayton, Lee and Moran, S. R., 1982. "Chronology of Late Wisconsin Glaciation in Middle North America". QUATERNARY SCIENCE REVIEWS, vol. 1.

Flint, Richard Foster, 1971. GLACIAL AND QUATERNARY GEOLOGY. New York: John Wiley and Sons.

Hole, F. D., 1976. SOILS OF WISCONSIN. Madison: University of Wisconsin Press.

Hough, J. L., 1963. "Prehistoric Great Lakes of North America". AMERICAN SCIENTIST, vol. 51.

Knox, J. C. and Mickelson, D. M., 1974. LATE QUATERNARY ENVIRONMENTS OF WISCONSIN, a field guide. American Quaternary Association. Wis. Geological and Natural History Survey.

Knox, J. C. and others, 1982. QUATERNARY HISTORY OF THE DRIFTLESS AREA, with special papers. Field trip guide book no. 5. Wis. Geological and Natural History Survey.

Matsch, Charles L., 1976. NORTH AMERICA AND THE GREAT ICE AGE. New York: McGraw-Hill Book Co.

Mickelson, David M., 1983. A GUIDE TO THE GLACIAL LANDSCAPES OF DANE COUNTY, WISCONSIN, with glacial geology map, scale 1:100,000. Field trip guide book no. 6. Wis. Geological and Natural History Survey.

Mickelson, David M. and others, 1983. LATE GLACIAL HISTORY AND ENVIRONMENTAL GEOLOGY OF SOUTHEASTERN WISCONSIN. Field trip guide book no. 7. Wis. Geological and Natural History Survey.

Mickelson, D. M., Clayton L., Baker, R. W., Mode, W. N., and Schneider, A. F., 1984. PLEISTOCENE STRATIGRAPHIC UNITS OF WISCONSIN. Misc. paper 84–1. Wis. Geological and Natural History Survey.

Schultz, Gwen, 1963. GLACIERS AND THE ICE AGE: EARTH AND ITS INHABITANTS DURING THE ICE AGE. New York: Holt, Rinehart, and Winston. Madison, Wis.: Reading Gems.

Schultz, Gwen, 1974. ICE AGE LOST. Garden City: Anchor Press/Doubleday. Madison, Wis.: Reading Gems.

Thwaites F. T., 1943. "Pleistocene of Part of Northeastern Wisconsin". BULL. OF THE GEOLOGICAL SOCIETY OF AMERICA, vol. 54.

Thwaites F. T. and Bertrand, K., 1957. "Pleistocene Geology of the Door Peninsula, Wisconsin". BULL. OF THE GEOLOGICAL SOCIETY OF AMERICA, vol. 68.

Thwaites, F. T., 1963. OUTLINE OF GLACIAL GEOLOGY. Self-published.

Maps

Topographic maps by U.S. Geological Survey. Show relief by contours, and surface features such as rivers, marshes, highways, trails, urban areas, buildings, and noteworthy sites. Scales—1:250,000, 1:62,500, 1:24,000, and (metric) 1:100,000. Index maps available. Wis. Geological and Natural History Survey.

The following are wall-size maps:

BEDROCK GEOLOGIC MAP OF WISCONSIN, Mudrey, M. G., Jr., Brown, B. A., and Greenberg, J. K., 1982. Scale 1:1,000,000. Wis. Geological and Natural History Survey.

BEDROCK GEOLOGY OF MARATHON COUNTY, WISCONSIN, LaBerge, Gene L. and Myers P. E., 1983. Scale 1:100,000. Wis. Geological and Natural History Survey.

BEDROCK GEOLOGY OF WISCONSIN, NORTHEAST SHEET, Greenberg, J. K. and Brown, B. A., 1984. Scale 1:250,000. Wis. Geological and Natural History Survey.

DEPTH TO BEDROCK IN WISCONSIN, Trotta, L. C. and Cotter, R. D., 1973. Scale 1:1,000,000. U.S. Geological Survey and Wis. Geological and Natural History Survey.

GLACIAL DEPOSITS OF WISCONSIN, 1976. Scale 1:500,000. Wis. Geological and Natural History Survey.

LANDFORMS OF WISCONSIN, 1972. Relief by David Woodward. Scale 1:100,000. Wis. Geological and Natural History Survey.

MADISON AREA LAKES, Patterson, D. L. and Czechanski, M. L., 1984. 1.5 in. to 4 mi. Wis. Geological and Natural History Survey.

PLEISTOCENE GEOLOGY OF BROWN COUNTY, WISCONSIN, Edward Need, 1983. Scale 1:100,000. Wis. Geological and Natural History Survey.

STATE OF WISCONSIN, TOPOGRAPHIC MAP, 1968. Scale 1:500,000; contour interval 200 feet. Wis. Geological and Natural History Survey.

STATE OF WISCONSIN, LANDFORMS MAP, 1968. Scale 1:500,000. Wis. Geological and Natural History Survey.

STATE OF WISCONSIN, HYDROGRAPHIC MAP (water features), 1968. Scale 1:500,000. Wis. Geological and Natural History Survey.

All maps listed here (and others, including page-size maps of various subjects) are obtainable from the Wisconsin Geological and Natural History Survey, University of Wisconsin, Madison. Its LIST OF PUBLICATIONS and index maps for selecting topographic maps are available upon request.

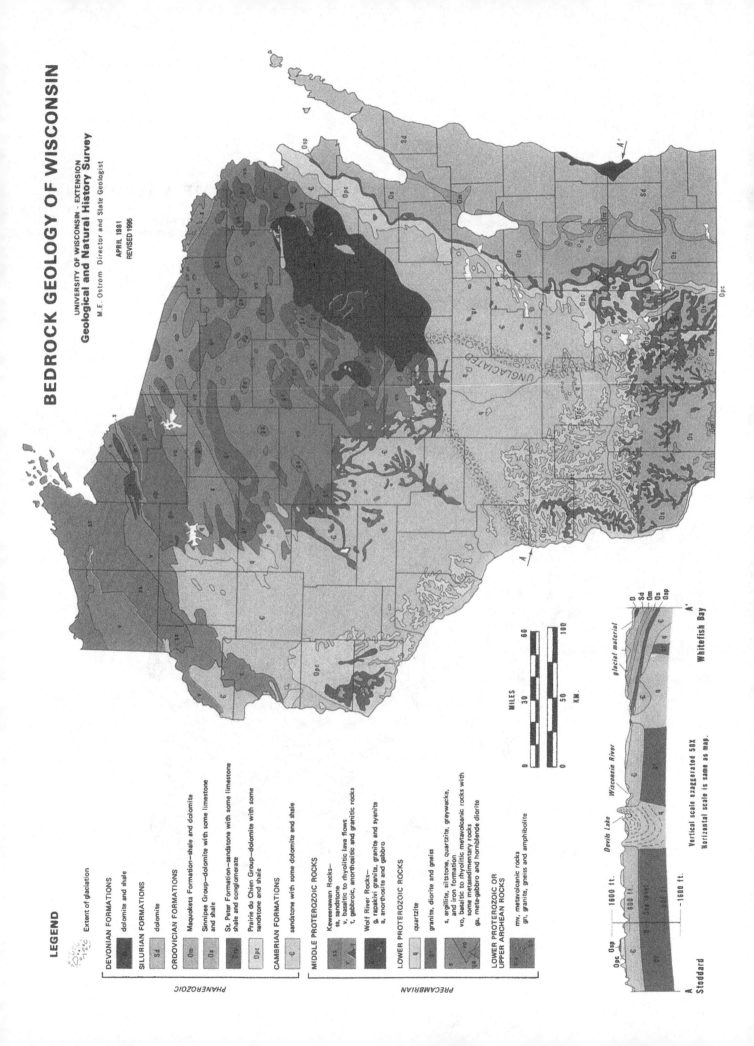

BEDROCK GEOLOGY OF WISCONSIN

UNIVERSITY OF WISCONSIN - EXTENSION
Geological and Natural History Survey
M.E. Ostrom Director and State Geologist

APRIL 1981
REVISED 1995

LEGEND

Extent of glaciation

PHANEROZOIC

DEVONIAN FORMATIONS
Sd dolomite and shale

SILURIAN FORMATIONS
Sd dolomite

ORDOVICIAN FORMATIONS
Om Maquoketa Formation—shale and dolomite

Os Sinnipee Group—dolomite with some limestone and shale

Osp St. Peter Formation—sandstone with some limestone shale and conglomerate

Opc Prairie du Chien Group—dolomite with some sandstone and shale

CAMBRIAN FORMATIONS
Є sandstone with some dolomite and shale

PRECAMBRIAN

MIDDLE PROTEROZOIC ROCKS
Keweenawan Rocks—
ss, sandstone
v, basaltic to rhyolitic lava flows
t, gabbroic, anorthositic and granitic rocks

Wolf River Rocks—
g, rapakivi granite, granite and syenite
a, anorthosite and gabbro

LOWER PROTEROZOIC ROCKS
q quartzite

granite, diorite and gneiss

s, argillite, siltstone, quartzite, graywacke, and iron formation
vo, basaltic to rhyolitic metavolcanic rocks with some metasedimentary rocks
ga, meta-gabbro and hornblende diorite

LOWER PROTEROZOIC OR UPPER ARCHEAN ROCKS
mv, metavolcanic rocks
gn, granite, gneiss and amphibolite

MILES
0   30   60

KM.
0   50   100

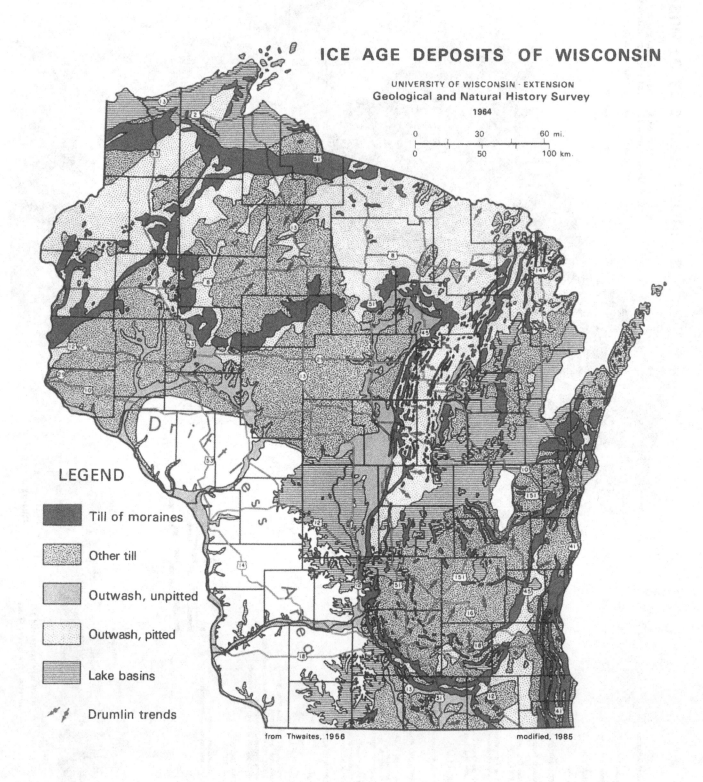

ICE AGE DEPOSITS OF WISCONSIN

UNIVERSITY OF WISCONSIN · EXTENSION
Geological and Natural History Survey
1964

LEGEND

Till of moraines

Other till

Outwash, unpitted

Outwash, pitted

Lake basins

Drumlin trends

from Thwaites, 1956

modified, 1985

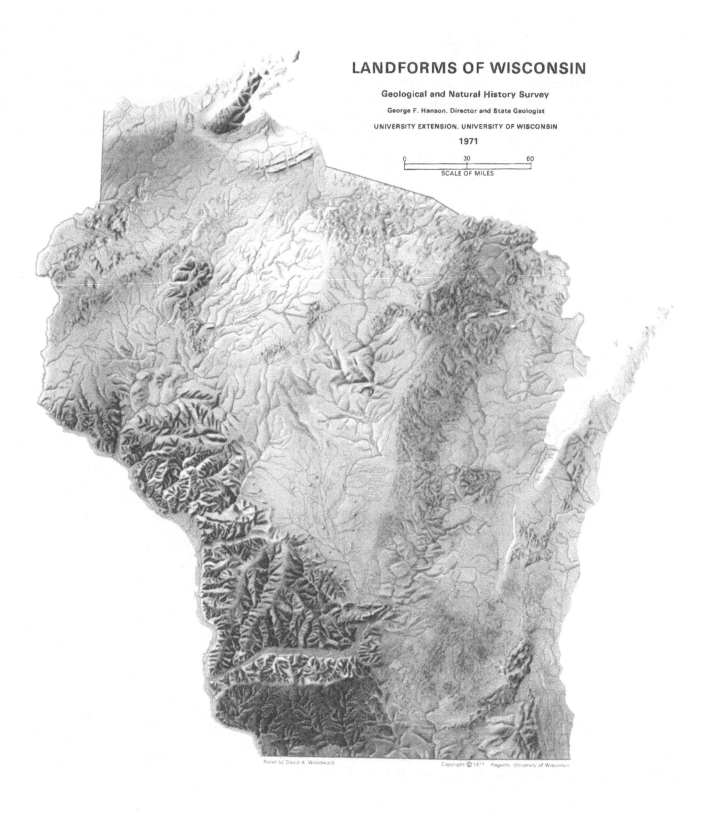

LANDFORMS OF WISCONSIN

Geological and Natural History Survey

George F. Hanson, Director and State Geologist

UNIVERSITY EXTENSION, UNIVERSITY OF WISCONSIN

1971

0   30   60

SCALE OF MILES

PLATE 4

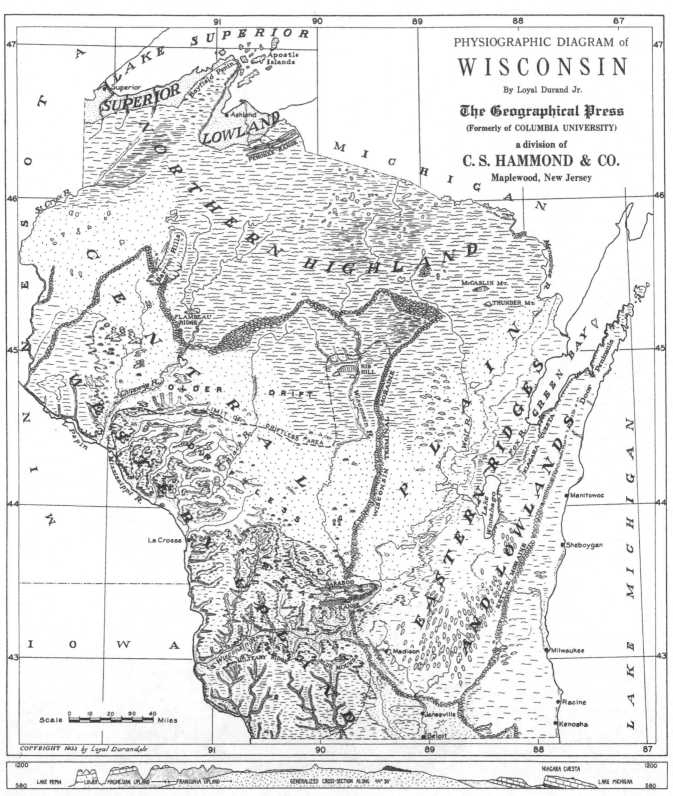

PHYSIOGRAPHIC DIAGRAM of
WISCONSIN
By Loyal Durand Jr.

The Geographical Press
(Formerly of COLUMBIA UNIVERSITY)
a division of
C.S. HAMMOND & CO.
Maplewood, New Jersey

COPYRIGHT 1933 by Loyal Durand, Jr.

Map courtesy of Hammond Inc., Maplewood, N.J.

PLATE 5

GEOLOGIC COLUMN OF PALEOZOIC ROCKS IN WISCONSIN

Time Units		Rock Units			Approximate Maximum Thickness in Feet
System	Series	Group	Formation	Member	
Devonian	Upper		Kenwood		235
	Middle		Milwaukee		
			Thiensville		
			Lake Church		
Silurian	Cayugan		Waubakee		620
	Niagaran		Racine		
			Manistique		
			Hendricks		
			Byron		
	Alexandrian		Mayville		
Ordovician	Cincinnatian		Neda		55
			Maquoketa		470
	Champlainian	Sinnipee	Galena		355
			Decorah		
			Platteville		
		Ancell	Glenwood		20
			St. Peter	Tonti	280
				Readstown	285
	Canadian	Prairie du Chien	Shakopee	Willow River	360
				New Richmond	
			Oneota		
Cambrian	St. Croixan	Trempealeau	Jordan	Coon Valley	110
				Sunset Point	
				Van Oser	
				Norwalk	
			St. Lawrence	Lodi — Black Earth	145
		Tunnel City	Lone Rock Mazomanie	Reno	170
				Tomah	
				Birkmose	
		Elk Mound	Wonewoc	Ironton	220
				Galesville	
			Bonneterre Eau Claire		250
			Mt. Simon		1,306

Meredith E. Ostrom, 1978
Thickness figures by Roger Peters, 1985

PLATE 6

REFERENCE MAP
OF WISCONSIN

Index

Lake St. Croix, 178
Lake Superior, 1, 139, 144, 145, 155, 165, 166, 185
    iron-formation, 33
    see Lake Superior syncline
Lake Superior Lowland, 47, 48, 49, 166
Lake Superior Sandstone, 35, 41, 49, 88
Lake Superior syncline, 35–42, 51
Lake Winnebago, 128, 131, 132, 168
Lake Wisconsin, 63
Land use, 1, 70–71, 73–74, 101, 103, 114, 115, 120, 124, 129, 134–137, 147, 171, 172, 180, 181, 182, 185, 186, 187
    see Environmental awareness
Lannon stone, 95, 134
Lapham, Increase, 163
Lapham Peak, 163
Laurentian, 2, 31
Lava, 28, 29, 37, 38, 40, 46–48, 50
    see Basalt, Rhyolite, Igneous rock, Keweenawan Area
Lead, 33, 60, 93, 120–127
Leland Natural Bridge, 63, 86, 111, 113
Limestone, 6, 31–32, 74, 77, 78, 134
    see Dolomite, Lower Magnesian Limestone
Linden, 122
Little Chute, 128
Lobes, glacial, 143, 144, 145, 155, 156, 158, 161
Lodi, 118
Lodi Siltstone, 86
Loess, 179–180
Lone Rock Formation, 85, 86, 110
Lower Magnesian Limestone, 88, 90
    cuesta, 100
    escarpment, 103
Lower Narrows (Baraboo River), 63, 67, 160
Lyndon Station, 103

Madeline Island, 45
    see Apostle Islands
Madison, 136, 148, 159, 170
    Four Lakes region, 87, 118, 121, 170
    Science Hall, 62
    Sunset Point, 87
Madison Sandstone, 87
Magma, 8, 27–28, 37–38, 40
    see Intrusions
Mammoth, woolly, 23, 182
Manitou Falls, 45, 47, 48
Manitowoc, 134
Manitowoc River, 133
Maquoketa Shale, 93, 94, 127, 129, 131, 132, 135
Marengo River, 32
Maribel, 136, 137
Marinette, 128
Marshes, 150, 152, 171, 175, 181, 182, 184, 185
    Central Sand Plain, 103
    Horicon Marsh, 128, 131
Marshfield, 57
Martin, Lawrence, preface, 27, 113, 114, 191
Mastodons, 23–24, 182, 183
Mauston, 103
Mayville, 131, 135
Mayville Formation, 95
Mazomanie Sandstone, 85, 86, 111, 113
McAllister, 89
McCaslin Mountain, 35, 55

Mellen, 41, 42, 45, 50, 52, 53
Meltwater, glacial, 174–179, 186
    see Outwash
Menasha, 128
Menominee Range, 33, 54–55
Menominee River, 1, 89
Menomonee Falls, 136
Menomonie, 158
Merrill, 57
Merrimac, 119
Metamorphic rocks, 6, 7, 8, 31–32, 146
Michigan, 1, 2, 40, 41, 45, 50, 57
    Antrim Shale, 96, 97
    Basin, 74–75, 94
    copper, 149
    diamonds, 148
    glaciation, 145, 155
    Gogebic Range, 51–52
    iron-formations, 33
    Lake Superior syncline, 36
    Menominee Range, 54, 55
    Paleozoic rocks, 82
    quartzites, 69
Midcontinent Gravity High, 41
Military Ridge, 14, 104, 120, 121
Military Road (U.S. Highway 18), 121, 131, 132
Mill Bluff, 109, 111, 112
    petroglyphs, 113
Millston, 106
Milwaukee, 96, 125, 134, 186
    sandstone buildings, 49
Milwaukee Formation, 96, 97
Milwaukee River, 96
Mineral Point, 122, 123, 124, 125
Mineral water, 74, 135, 136, 137
Mining, 132, 133, 135
    Baraboo Hills, 67
    Black River Falls, 59
    Central Shield Area, 60
    Gogebic Range, 52
    Ironton, 84
    Menominee Range, 55
    southwest lead-zinc district, 120–127
Minnesota, 2, 40, 41, 43, 45, 50, 75, 86, 119, 149
    Dresbach, 83
    Duluth, 39, 42
    Franconia, 84
    glaciation, 144, 155
    iron-formations, 33
    Jordan Sandstone, 86
    Lake Superior syncline, 36, 37
    Paleozoic rocks, 82
    pipestone, 56, 70
    quartzites, 69–70
    Shakopee, 89
    St. Lawrence Dolomite, 86
    St. Peter Sandstone, 90
Mississippi River, 75, 84, 105, 114, 118, 119, 120, 121, 123, 125, 128, 129, 178, 179
Monadnocks, 32, 58–62, 81, 101
    see Baraboo Hills
Montello, 29, 61, 62, 146
Montfort, 120, 122
Montreal River, 1, 45, 51
Moraines, 150, 153, 155, 159, 161, 170, 175, 186, 187, 188
    end moraine, 150
    ground moraine, 151
    interlobate moraine, 151, 158
    recessional moraine, 151
    terminal moraine, 151

Mosinee, 58
Mosinee Flowage, 57
Mosinee Hill(s), 59
Mounds, 62, 104, 106, 107, 114
    see Blue Mound, Central Sand Plain, Indians, Iron Mound, monadnocks, outliers, Platte Mound
Mountains, 30, 31, 34, 55
    Flambeau Mountain, 56
    McCaslin Mountain, 55
    Rib Mountain, 58–59
    Thunder Mountain, 55
    Wildcat Mountain, 111
Mount Ashwabay, 165
Mount Hope, 120
Mount Horeb, 120
Mount Simon Sandstone, 82, 83, 90, 101, 135
Mount Whittlesey, 45, 53

Natural bridges
    Leland, 63, 86, 111, 113
    Rockbridge, 113
Necedah, 62, 103
Necedah Mound, 62
Neda Formation, 94, 135
Neenah, 128
Neillsville, 103
    Black River Falls—Neillsville Area, 44, 59
    mounds, 104
Nekoosa, 58, 167
New Diggings, 122
New Richmond Sandstone, 89
Niagara Cuesta, 100, 105, 119, 127, 129–134, 146, 159, 163
Niagara dolomite, 18, 88, 94, 95, 99, 130, 131, 134
Niagara Escarpment, 102, 104, 128, 129, 131–133
Niagara Fault, 54, 55
Nonesuch Shale, 40, 48
North Freedom, 67
Norwalk Sandstone, 86, 87

Observatory Hill
    Marquette county, 61
    Madison campus, 147
Oconomowoc, 128
Oconto, 128, 136
Oconto Falls, 89
Oil, 48, 51, 135
Old drift, 158, 162, 181
Oneota Dolomite, 89
Oolites, 89, 94, 135
Ordovician Period, 10, 12–15
    strata, 88–94
Orogenies, 30
    Algoman, 31
    Penokean, 31, 34, 51, 59, 63, 70
Oronto Group, 35, 36, 40, 41
Osceola, 136
Oshkosh, 128, 136
Outliers, 17, 87, 94, 100, 104, 106–114, 118, 131, 132, 133
Outwash, glacial, 153, 160, 170, 174–179
    see Pitted outwash
Outwash plains, 175, 176, 177, 186